Readers Praise the
Math You Can Play Series

These games are great for using and practicing maths skills in a context in which there is some real motivation to do so. I love how they provide opportunities to explore a wide variety of approaches, including number bonds and logical thinking.

The games use very simple materials, mostly cards and dice, and the few boards that are needed are provided. Each game also has tips on how to customise or extend it—maybe for players of different abilities, or non-competitive options. My children are always pleased, even excited, when I suggest one of these games. Sometimes they even ask to play them unprompted!

—Miranda Jubb, Amazon reviewer

I have played several of these games with my son, and each one was met with delight on his part and the sharing of delightful conversation about numbers and thinking between us.

I love what Gaskins has to say about working *with* your children as opposed to simply assigning them work to do. This sums up the philosophy that I try to keep forefront in our home, and it's the thing that makes these books such a valuable addition to our library of educational resources.

—Amy, Hope Is the Word blog

Wonderful games for elementary students. The author includes a link for printable game boards, ensuring that I don't spend more time making games than playing them. Variations for each game = *so* many ways to explore numbers. You will love this book.

—Marisa, Amazon reviewer

Multiplication & Fractions

Math Games for Tough Topics
Second to Sixth Grade

Denise Gaskins

TABLETOP ACADEMY PRESS

Print version 1.01
Many sections of this book were originally published on the Let's Play Math blog.
denisegaskins.com

Tabletop Academy Press, Blue Mound, IL, USA
tabletopacademy.net

ISBN: 978-1-892083-23-4
Library of Congress Control Number: 2016918012

Cover photo by Karen Struthers via Dreamstime:
dreamstime.com/royalty-free-stock-photos-man-playing-cards-his-children-image1567678

Fractions display by Miss K Primary (CC-BY 2.0):
flickr.com/photos/misskprimary/1037165017

Riffle shuffle photo by Johnny Blood (CC-BY-SA 2.0):
commons.wikimedia.org/wiki/File:Riffle_shuffle.jpg

Author photo by Mat Gaskins.
matgaskins.com

Contents

Resources and References

Preface to the *Math You Can Play* Series

THE PLAYFUL, PUZZLE-SOLVING SIDE OF math has always attracted me. In elementary school, calculations were a tedious chore, but word problems provided the opportunity to try out my deductive powers. High school algebra and geometry were exercises in logical reasoning, and college physics was one story problem after another—great fun!

As my children grew, I wanted to share this sort of mathematical play with them, but the mundane busyness of everyday life kept pushing aside my good intentions. Determined to make it happen, I found a way to defeat procrastination: invite friends to bring their kids over for a math playdate. We grappled with problems, solved puzzles, and shared games. Skeptical at first, the kids soon looked forward to math club. When that gang grew up and moved on, their younger siblings came to play, and others after them. Sometimes we met weekly, sometimes monthly or just off and on. At our house, at the library, in the park—more than twenty years of playing math with kids.

Now I've gathered our favorite math club games into these *Math You Can Play* books. They are simple to learn, easy to set up, and quick to play, so even the busiest parents can build their children's mental math skills and promote logical thinking.

I hope you enjoy these games as much as we have. If you have any questions, I would love to hear from you.

—DENISE GASKINS
LETSPLAYMATH@GMAIL.COM

P.S.: If you've read the *Math You Can Play Series* books in order, you will notice that I repeat myself in Sections I and III. I'm including the setup information and math teaching tips in each book to make sure they can all stand on their own.

Acknowledgements

No man is an island, entire of itself.
Every man is a piece of the continent, a part of the main.
If a clod be washed away by the sea,
Europe is the less, as well as if a promontory were.

—John Donne

Neither an island nor a promontory—I am that little clod supported by a continent of family, friends, and online acquaintances whose help and encouragement have made my math books possible. I can't express the debt I owe to my husband David, whose patience stretches far beyond what I deserve, or to our children, who taught me so much over the many years of homeschooling.

Thank you to Marilyn Kok and Sue Kunzeman, who brought their children to that first math playdate—and kept bringing them back. Special thanks to the many math club kids who joined in the activities and tested out the games. Thank you to John Golden, whose Math Hombre blog inspires me to think deeply about math games, and to Sue VanHattum, whose work on her own book, *Playing With Math: Stories from Math Circles, Homeschoolers, and Passionate Teachers*, convinced me to bring my books back into print. To my many thousands of book and blog readers, your comments have kept me going. To my fellow math bloggers, I've learned so much from you all!

Fervent thanks to my beta readers: Becky, Emily, Jennifer, Katie, Laura, Marcia, Maria, Marisa, Nicky, Roxana, Sharon, Siobhan, and

Sue. And unending gratitude to my friend and editor Robin Netherton. Whatever mistakes remain are due to my continual tinkering with the text after it left her hands.

Section I

A Strategy for Learning

Prof. Triangleman's Abbreviated List of Standards for Mathematical Practice

PTALSMP 1: Ask questions.

Ask why. Ask how. Ask whether your answer is right. Ask whether it makes sense. Ask what assumptions you have made, and whether an alternate set of assumptions might be warranted. Ask what if. Ask what if not.

PTALSMP 2: Play.

See what happens if you carry out the computation you have in mind, even if you are not sure it's the right one. See what happens if you do it the other way around. Try to think like someone else would think. Tweak and see what happens.

PTALSMP 3: Argue.

Say why you think you are right. Say why you might be wrong. Try to understand how someone else sees things, and say why you think their perspective may be valid. Do not accept what others say is so, but listen carefully to it so that you can decide whether it is.

PTALSMP 4: Connect.

Ask how this thing is like other things. Try your ideas out on a new problem. Ask whether and how these ideas apply to other situations. Look for similarities and differences. Seek out the boundaries and limitations of your techniques.

—Christopher Danielson

Practice applying Professor Triangleman's Standards
to the puzzle below. Which one doesn't belong?
Can you find more than one answer?

There should be no element of slavery in learning.
Enforced exercise does no harm to the body, but enforced
learning will not stay in the mind. So avoid compulsion,
and let your children's lessons take the form of play.

—PLATO

Introduction: How to Use This Book

IF A PERFECT TEACHER DEVELOPED the ideal teaching strategy, what would it be like?

- ♦ An ideal teaching strategy would have to be flexible, working in a variety of situations with students of all ages.

- ♦ It would promote true understanding and reasoning skills, not mere regurgitation of facts.

- ♦ It would prepare children to learn on their own.

- ♦ Surely the ideal teaching strategy would be enjoyable, perhaps even so much fun that the students don't realize they are learning.

- ♦ And it would be simple enough that imperfect teachers could use it, too.

This is idle speculation, of course. No teaching strategy works with every student in every subject. But for math, at least, there is a wonderful way to stimulate our children's number skills and encourage them to think: we can play games.

Math games push students to develop a creatively logical approach to solving problems. When children play games, they build reasoning skills that will help them throughout their lives. In the stress-free struggle of a game, players learn to analyze situations and draw conclusions. They must consider their options, change their plans in reaction to the other player's moves, and look for the less obvious solutions in order to outwit their opponents.

Even more important, games help children learn to enjoy the challenge of thinking hard. Children willingly practice far more arithmetic than they would suffer through on a workbook page. Their vocabulary grows as they discuss options and strategies with their fellow players. Because their attention is focused on their next move, they don't notice how much they are learning.

And games are good medicine for math anxiety. Everyone knows it takes time to master the fine points of a game, so children can make mistakes or "get stuck" without losing face.

If your child feels discouraged or has an "I can't do it" attitude toward math, try taking him off the textbooks for a while. Feed him a strict diet of games, and his eyes will soon regain their sparkle. Children love beating a parent at a math game. And if you're like me, your kids will beat you more often than you'll want to admit.

Math You Can Play

Clear off a table, find a deck of cards, and you're ready to enjoy some math. Most of the games in this book take only a few minutes to play, so they fit into your most hectic days.

In three decades of teaching, I've noticed that flexibility with mental calculation is one of the best predictors of success in high school math and beyond. So the *Math You Can Play* games will stretch your children's ability to manipulate numbers in their heads. But unlike the typical "computerized flash card" games online, most of these games will also encourage your children to think strategically, to compare different options in choosing their moves.

"Be careful! There are a lot of useless games out there," says math professor and blogger John Golden. "Look for problem solving, the need for strategy, and math content.

"The best games offer equal opportunity (or nearly so) to all your students. Games that require computational speed to be successful will disenfranchise instead of engage your students who need the game the most."

Each book in the *Math You Can Play* series features twenty or more of my favorite math games, offering a variety of challenges for all ages. If you are a parent, these games provide opportunities to enjoy quality time with your children. If you are a classroom teacher, use the games as warm-ups and learning center activities or for a relaxing review day at the end of a term. If you are a tutor or homeschooler, make games a regular feature in your lesson plans to build your students' mental math skills.

Know that my division of these games by grade level is inherently arbitrary. Children may eagerly play a game with advanced concepts if the fun of the challenge outweighs the work involved. Second- or third-grade students can enjoy some of the games in the prealgebra book. On the other hand, don't worry that a game is too easy for your students, as long as they find it interesting. Even college students will enjoy a round of Farkle (in the addition book) or Wild and Crazy Eights (a childhood classic from the counting book). An easy game lets the players focus most of their attention on the logic of strategy.

As Peggy Kaye, author of *Games for Math*, writes: "Children learn more math and enjoy math more if they play games that are a little too easy rather than a little too hard."

Games give children a meaningful context in which to ponder and manipulate numbers, shapes, and patterns, so they help players of all skill levels learn together. As children play, they exchange ideas and insights.

"Games can allow children to operate at different levels of thinking and to learn from each other," says education researcher Jenni Way. "In a group of children playing a game, one child might be encountering a concept for the first time, another may be developing his/her understanding of the concept, a third consolidating previously learned concepts."

Talk with Your Kids

The modern world is a slave to busyness. Marketers tempt well-intentioned parents with toys and apps that claim to build academic skills while they keep our children occupied. Homeschoolers dream of finding a curriculum that will let the kids teach themselves. And even the most attentive teachers may hope that game time will give them a chance to correct papers or catch up on lesson plans.

Be warned: although children can play these games on their own, they learn much more when we adults play along.

When adults play the game, we reinforce the value of mathematical play. By giving up our time, we prove that we consider this just as important as *[insert whatever we would have been doing]*. If the game is worthy of our attention, then it becomes more attractive to our children.

Also, as we watch our kids' responses and listen to their comments during a game, we discover what they understand about math. Where do they get confused? What do they do when they're stuck? Can they use the number relationships they've mastered to figure out something they don't know? How easily do they give up?

"Language should be part of the activity," says math teacher and author Claudia Zaslavsky. "*Talk* while you and your child are playing games. Ask questions that encourage your child to describe her actions and explain her conclusions."

Real education, learning that sticks for a lifetime, comes through person-to-person interactions. Our children absorb more from the give and take of simple discussion with an adult than from even the best workbook or teaching video.

If you're not sure how to start a conversation about math, browse the stories at Christopher Danielson's Talking Math with Your Kids blog.[†]

As homeschooler Lucinda Leo explains, "With any curriculum there is the temptation to leave a child to get on with the set number of pages while you get on with something else. My long-term goal is for my kids to be independent learners, but the best way for that to happen is for me to be by their side now, enjoying puzzles and stories, asking good questions and modelling creative problem-solving strategies."

And playing math games.

Mixing It Up

Games evolve as they move from one person to another. Where possible, I have credited each game's inventor and told a bit of its history. But some games have been around so long they are impossible for me to trace. Many are adapted from traditional childhood favorites. For example, I was playing Tens Concentration with my math club kids years before I read about it in Constance Kamii's *Young Children Reinvent Arithmetic*. Similarly, an uncountable number of parents and teachers have played Math War with their students; a few of my variations are original, but the underlying idea is far from new.

Sometimes, as in the Math War variations, the basic rules of play stay the same, making the new games easy for children to learn. Likewise, the multiplication classic Product Game can be modified to practice decimal or fraction multiplication without changing the rules, just by switching the game board.

In other cases, we change the rules themselves to create a completely new game. Consider the lineage of Forty-Niners, featured in the *Math You Can Play* addition book. First someone invented dice, and generations of players created a multitude of folk games, culminating in Pig. Using cards instead of dice and adding a Wild West

---

† *I'll refer to dozens of blogs, websites, and other resources throughout this book. All of these (and more) are listed in the appendix "Quotes and Reference Links" on page 134.*

theme, James Ernest created the Gold Digger version and gave it away at his website. Teachers wanted their students to practice with bigger numbers, so they tried a regular deck of playing cards, and the game became Stop or Dare at the Nrich website. For my variation, I increased the risk level by turning all the face cards into bandits and adding the jokers as claim jumpers.

Game rules are a social convention, easy to change by agreement among the players. Feel free to invent your own rules, and encourage your children to modify the games as they play. For instance:

- ♦ Can you make the game easier, so young children can play? Or harder, to challenge adults?

- ♦ What would happen if you changed the number of moves? Or the number of cards you draw, or how many dice you throw?

- ♦ Can you invent a story to explain the game—like James Ernest did with Gold Digger—or tie it to a favorite book?

- ♦ If the game uses cards, can you figure out a way to play it with dice or dominoes? Or transfer it to a game board?

- ♦ If the game uses a number chart, could you play it on a clock or calendar instead? Or is there a way to use money in the game?

- ♦ Or can you change it into a whole-body action game? Perhaps using sidewalk chalk?

As children tinker with the game, they will be prompted to think more deeply about the math behind it.

Unschooling advocate Pam Sorooshian explains the connection between games and math this way:

Mathematicians don't sit around doing the kind of math that you learned in school. What they do is "play around" with number games, spatial puzzles, strategy, and logic.

They don't just play the same old games, though. They change the rules a little, and then they look at how the game changes.

So, when you play games, you are doing exactly what mathematicians really do—if you fool with the games a bit, experiment, see how the play changes if you change a rule here and there. Oh, and when you make up games and they flop, be sure to examine why they flop—that is a big huge part of what mathematicians do, too.

Finally, although the point of these games is for children to practice mental math, please don't treat them as worksheets in disguise. A game should be voluntary and fun. No matter how good it sounds to you, if a game doesn't interest your kids, put it away. You can always try another one tomorrow.

You will know when you find the right game because your children will wear you out wanting to play it again and again and again.

1	2	3	4	5	6	7	8	9	10
11	12	13	14	15	16	17	18	19	20
21	22	23	24	25	26	27	28	29	30
31	32	33	34	35	36	37	38	39	40
41	42	43	44	45	46	47	48	49	50
51	52	53	54	55	56	57	58	59	60
61	62	63	64	65	66	67	68	69	70
71	72	73	74	75	76	77	78	79	80
81	82	83	84	85	86	87	88	89	90
91	92	93	94	95	96	97	98	99	100

You can play many games on a hundred chart.
The *Math You Can Play* printable companion files
include 0–99 charts, too , for those who prefer that format.

We do not stop playing because we grow old. We grow old because we stop playing.

—Anonymous

Gather Your Game Supplies

I HAVE A LIMITED AMOUNT of free time, and I don't want to spend it cutting out specialized game pieces or cards. A few games require printable cards or game boards, but most of the games in the *Math You Can Play* series use basic items you already have, such as playing cards and dice.

A Deck of Math Cards

Whenever a game calls for playing cards, I use an international standard poker- or bridge-style deck (or *pack;* the terms are interchangeable). There are fifty-two cards in four suits—spades (the pointy black shape), hearts, clubs (the clover shape), and diamonds—with thirteen cards per suit. The number cards range from the ace to ten, and each suit has three face cards called jack, queen, and king. Your deck may have one or two additional cards called jokers, which are not officially part of the deck but may be used for some games.

Math cards are simply the forty number cards (ace through ten in all four suits) from a standard deck. The ace counts as "one" in all math card games. Some game variations call for using the face cards as higher numbers: jack = 11, queen = 12, and king = 13. In a few games, we use the queens as zeros, because the Q is round enough for pretend.

Other types of card decks may work as well, so experiment with whatever you have on hand. Uno cards are numbered zero to nine, Phase 10 cards have one to twelve, and Rook cards go from one to fourteen. Rummikub tiles use the numbers one to thirteen. Most of the games in this book could be adapted to use any of these.

Game Boards

Many games use graph paper or a hundred chart, which you can find online. For most other games, hand-drawn boards work fine. One reason Tic-Tac-Toe is a perennial favorite is that children can draw the board whenever they want to play.

I've created two free PDF packets of charts and game boards—the *Number Game Printables Pack* and *Multiplication & Fraction Printables*—which you can download from my blog. You may reproduce these for use within your own family, classroom, or homeschool group.

To save paper, you may wish to reuse game boards. Print the game board on cardstock and laminate it—I love my laminator!—or slip the

You can find the free printable companion files on my Tabletop Academy Press publishing website:
tabletopacademy.net/free-printables

printed game board into a clear (not frosted) page protector, adding a few sheets of card stock or the back of an old notebook for stiffness. Then your children can mark moves with dry-erase markers and wipe them clean with an old, dry cloth.

A few games call for larger homemade boards. For example, Dinosaur Race (in the counting book) needs a simple track, twelve to twenty spaces long, and each space needs to be large enough for a couple of toy dinosaurs or other small figures. An open manila file folder can serve as a sturdy foundation on which to draw or paste the board, convenient for playing and easy to store. And if you keep a stack of blank manila folders freely available, your children will enjoy making up their own board games.

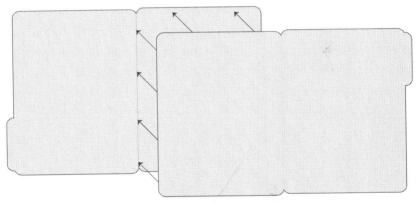

Glue two manila folders together to make
an even bigger game board.

Other Bits and Pieces

Many games call for small toy figures or other items to mark the players' position or moves. If two different types of tokens are needed, you may borrow the pieces from a checkers game. Or use pennies and nickels, milk jug lids in different colors, dried pinto and navy beans, or inexpensive acrylic stones from the craft section of your local department store. Or let players create their own game pieces by cutting small shapes from colored construction paper, decorating if desired.

For some games, tokens may be replaced by colored erasable markers on a laminated game board. The colors must be different enough to easily distinguish each player's moves, or one player can mark X and the other mark O. Some of the colored dry-erase markers leave stains, but you can wash off stubborn marks with rubbing alcohol or window cleaner.

When a game calls for dice, I have in mind the standard six-sided cubes with dots marking the numbers one through six. Most games only need one or two dice, but Farkle requires six. In many of the games, you may substitute higher-numbered dice for a greater challenge. And children enjoy using novelty dice when making up their own games.

A few games call for either a double-six or double-nine set of dominoes. If you are buying these, I recommend getting the larger set. You can always set aside the higher-numbered tiles when playing with young children.

Ready to Play?

If you want to put together a game box to keep all your supplies in one place, you will need:

- ◆ standard playing cards (two or more decks)
- ◆ pencils or pens
- ◆ colored felt-tip markers or colored pencils
- ◆ erasable markers for use on laminated game boards, and a cloth or paper towel for wiping them clean
- ◆ blank paper
- ◆ at least two kinds of tokens
- ◆ dice
- ◆ dominoes

♦ graph paper in assorted sizes[†]

♦ a couple of hundred charts (or 0–99 charts)[‡]

Try to let children learn by playing. Explain the rules as simply as possible and get right into the fun. You can add details, exceptions, and special situations as they come up during play or before starting future games. At our house, we play a few practice rounds first, and I make sure all the rules have been explained before we keep score.

Card games have a traditional ethic that guides players in choosing who gets to deal, who goes first, what to do if something goes wrong in the deal or during play, and more. If you are unsure about questions of this sort, read the appendix "Game-Playing Basics" on page 123.

Many of the game listings include suggestions for *house rules,* which are optional modifications of the game. The way a game is played varies from one place to another, and only a few tournament-style games have an official governing body to set the rules. If you're not playing in an official competition, then everything is negotiable. Players should make sure they agree on the rules before starting to play.

[†]   *incompetech.com/graphpaper*

[‡]   *themathworksheetsite.com/h_chart.html*

Multiplication and Fraction Games

Mathematical Models

"Mastery" in this context means not just being able to perform calculations with fluency. It is also important to have a good conceptual understanding of numbers, arithmetic, and reasoning, particularly in the context of real-world applications.

Everyday math involves a whole lot more than rote learning of a few facts. You can learn to calculate with numbers without any real understanding of the underlying concepts.

But applying arithmetic to things in the world, to quantities, and understanding the relationships between those quantities, requires considerable understanding of those underlying concepts.

—Keith Devlin

What Is a Math Model?

Teachers and parents know that every elementary math student needs to master the number facts. To memorize so many details can seem like an unending task. Too often, we adults are tempted to stress the rote aspect of such memory work, which makes our children lose their focus on what the numbers mean. But if we concentrate first on learning and using *math models*—physical or pictorial representations that help students make sense of mathematical concepts—we give our kids a strong foundation for middle school math.

Models give us a way to form and manipulate an image of an abstract concept, such as a fraction. When teaching young students, we act out ideas using blocks, cookies, or pieces of construction paper. Older children firm up their understanding by drawing pictures. As kids grow up and face more abstract, numbers-only problems, these pictures remain in their minds, an always-ready tool to help them reason their way through multiplication or fraction problems.

The reason we teach more than one model is that none of them can fit every type of problem. They are thinking tools, and every tool has its limit. Keep in mind semantics expert Samuel Hayakawa's first principle: "The symbol is *not* the thing symbolized; the word is *not* the

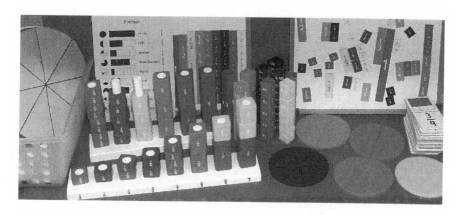

Your child's math curriculum may offer a variety of physical or pictorial models to promote comprehension.

thing; the map is *not* the territory it stands for." And the math model is not the number itself.

You can work with your children to create a deck of math model cards, sketching the pictures on index cards or on a stack of old business cards with blank backs. Each deck should include 10–15 sets, or *books*. Take your time, making just two or three books each day while talking about real-life situations the models might symbolize. When you think the deck is finished, lay the cards out on the table in sets, to make sure each book has all its members.

Or take the quick route: download my free *Multiplication & Fraction Printables* file, which includes two decks of math model cards along with all the game boards you need for any game in this book. Print the pages on card stock, and laminate for durability if desired.[†]

Multiplication Model Cards

Your card deck will not include every math fact in the times tables. But it should offer enough variety to cement the most common multiplication models in your children's minds. My multiplication card deck includes products from 2 × 2 to 6 × 6. Each book consists of four cards—the multiplication equation and the following three pictures:

(1) Set Model: "___ sets of ___ objects per set"

This model represents discrete (countable) items collected into groups: apples per basket, pennies per dime, or cookies per plate. The set model connects the concepts of multiplication and addition, so it is the most common way of introducing multiplication in elementary school textbooks.

(2) Bar (Measurement) Model: "___ units of ___ parts per unit"

This model represents continuous quantities measured out in parts: inches per foot, cups per recipe, dollars per pound, or

[†] *tabletopacademy.net/free-printables*

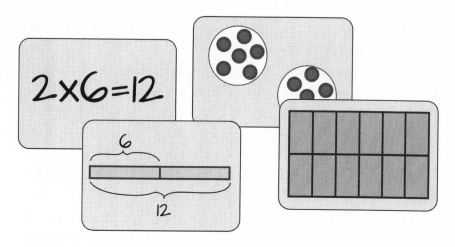

One book of multiplication model cards.

spaces per jump on a number line. The measurement model can also include scaling, stretching, or shrinking something from its original size, which makes it useful when thinking about fractions.

"Array"
(3) Rectangular Model: "___ rows of ___ items per row"

In early elementary math, this model represents an array of discrete items: chairs per row, buttons per column, or soldiers on parade. As students grow, however, the model expands to include continuous rectangular area. At its most mature, this model becomes the basis for many topics in high school math and beyond, including integral calculus. Because of its flexibility, the rectangular model is the most important one for our children to master.

Notice that in each model, the two numbers of a multiplication problem have different roles. One number is a *scale factor*, which tells you how many sets, units, or rows you are talking about, while the other number is a *this-per-that ratio*.

In addition and subtraction, numbers count how much stuff you have. If you get more stuff, the numbers get bigger. If you lose some of the stuff, the numbers get smaller. Numbers measure the amount of cookies, horses, dollars, gasoline, or whatever.

The this-per-that ratio is not a counting number, but something new. Something alien, completely abstract. It doesn't count the number of dollars or measure the volume of gasoline but tells the relationship between them, the dollars per gallon, *which stays the same whether you buy a lot or a little.* A ratio is a relationship number.

This is why you may hear math specialists like Stanford's Keith Devlin say, "Multiplication is not repeated addition." Children can use addition to solve whole-number multiplication puzzles, but that will not get them very far. Until they wrestle with and come to understand the concept of ratio, our students can never master multiplication.

Fraction Model Cards

When children can visualize the simple fractions included in the math models card deck, they can apply that knowledge to any fraction they meet. My fraction deck includes halves, thirds, fourths, and sixths. Each book consists of four cards—the fraction itself and the following three pictures:

(1) Round Food Model: Slices of pizza or pie

This is an easy way for most people to visualize fractions. We are used to cutting up pizza or pumpkin pie into various-sized slices, depending on the number of people we are serving. And because children usually like food, this makes fractions seem less threatening.

You can cut apart other shapes, too. My children's first experiences with fractions came in deciding how they wanted their grilled cheese sandwiches cut. You can turn this into a fun puzzle: draw a square and then draw lines that slice it vertically, horizontally, and on both diagonals. How many different ways can your child color exactly half the square?

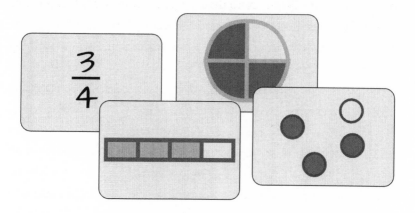

One book of fraction model cards.

(2) Bar (Measurement) Model: A ruler or measuring tape

Just as the number line is our most useful model for working with numbers in general, so the bar (which is a thickened portion of the number line) will be our most flexible model for fractions. Cut the bar vertically, like the marks on a ruler.

The bar model is easy to draw and divide. And it connects conceptually to the rectangle model of multiplication, which will help when students need to make sense of multiplying fractions.[†]

(3) Set Model: Parts of a group

This is the least natural way for many children to understand fractions. If these fractions seem awkward to your kids, keep coming back to them until they make sense. "Parts of a group" will show up later in math topics like percentages, probability, and statistics.

When using these models, remember that a fraction is a number, not a recipe for action. The fraction ¾ does not mean "Cut a pizza into four pieces, and then keep three of them." The fraction ¾ names

† *denisegaskins.com/2015/12/17/understanding-math-fraction-multiplication*

a certain amount of stuff, more than a half but not as much as a whole thing. When our students are learning fractions, we do cut up models to help them understand, but the fractions themselves are simply numbers.

Before a fraction can have any meaning, we must define something as "one whole unit." Of course, this is true of all numbers, not just fractions. Before *seven* can have any meaning, we must define *one*. Are we talking about seven miles or seven bags of rice or seven spaces on the number line? Natural numbers are defined in terms of whatever *one* is, so fractions are no different from other numbers in this respect.[†]

Because a fraction is a number, it can be added to other numbers (or subtracted, multiplied, etc.), and it has to obey the distributive property and all the other standard rules for numbers. We need two digits (plus a bar) to write a fraction like ⅛, just as it takes two digits to write the number 18—but, like eighteen, the fraction is a single number that names a certain amount of whatever we are counting or measuring.

On the Other Hand

The term *mathematical model* also can describe equations or other mathematical expressions that help us understand something in the real world. Biologists make models of population growth. Engineers model the stresses within a mechanical system. Medical researchers model the concentration of drugs in the bloodstream. To learn more about this type of modeling, visit the Mathematical Models page at Maths Is Fun.[‡]

---

† *See Christopher Danielson's short video "One is One ... Or Is It?" on YouTube:* *youtu.be/EtclcWGG7WQ*
‡ *mathsisfun.com/algebra/mathematical-models.html*

Twelve Cards

MATH CONCEPTS: multiplication or fraction models.

PLAYERS: two or more.

EQUIPMENT: one deck of math model cards.

How to Play

The first player shuffles the deck and then turns up the top twelve cards, placing them face up in a 3 × 4 array: three rows with four cards in each row. That player removes any pairs of cards that show the same product or fraction, keeping them in a personal score stash on the table.

After claiming all visible pairs, the first player passes the deck to the next player. Each player in turn deals out enough cards to fill in the empty spots in the array, captures any matching pairs, and passes the deck on.

On rare occasions, you'll get a dead hand where none of the cards you deal out will match. In that case, deal another set of twelve cards on top of the first. Claim the visible pairs, and also claim any bonus pairs that show up as you reveal the cards underneath.

Cards may only be taken in pairs, so if three cards match, you must leave one of them for the next turn. But if all four cards of a set are showing, you may take both pairs.

The game ends when there aren't enough cards left to fill the holes in the array. Whoever has collected the most cards wins the game.

Variations

HOUSE RULE: If you're playing with the fraction models, how will you handle equivalent fractions? Decide in advance whether you will allow a card labeled 3/6 to match with a picture of 2/4, since both of them are worth half of one whole thing.

NINE CARDS: If your children take too long inspecting their twelve-card tableau, simplify the game by using a 3 × 3 array. Dead hands are more common with the smaller array, but using fewer cards will help. Remove three or four books from your deck before playing.

History

I adapted this game from the number bonds game Nine Cards in Constance Kamii's *Young Children Continue to Reinvent Arithmetic: 2nd Grade.*

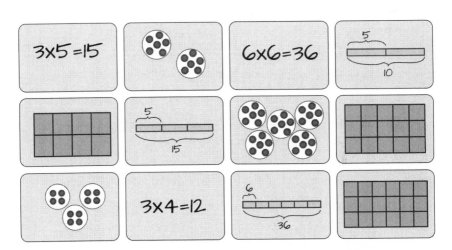

Can you find all four matching pairs of cards? When three cards match, you may choose any two of them and leave the third.

Go Fish

MATH CONCEPTS: multiplication or fraction models.

PLAYERS: two or more.

EQUIPMENT: one deck of math model cards.

How to Play

Deal seven cards to each player, or five cards if you have four or more players. Put the remaining cards face down in the center of the table and spread them to make a roughly circular "fishing pond."

Players look at their own hands and lay down pairs that show the same product or fraction. Each player creates a personal score pile, or "fish basket."

On your turn, you may ask one other player, "Do you have a _____?" The blank is for the multiplication expression or fraction that matches a card in your hand. For instance, if you have a 3 × 4 card, you might ask, "Do you have a 3 × 4 = 12?" With more than two players, the request must be addressed to a specific person. You may ask, "George, do you have a ⅔?" but it is illegal to say, "Does anyone have a ⅔?"

If George has the card you want, he must give it to you. If not, he says, "Go Fish." Then you draw any card you wish from the fishing pond. If you draw a card that matches any card in your hand, add that pair to your fish basket. Otherwise, add the card to your hand.

If the fishing pond goes dry, every player who has two or more cards in hand places one card face down to replenish the stock. Mix these cards thoroughly before continuing the game.

The game is over when one player runs out of cards. The other players throw their remaining cards into the fishing pond. (Those fish were too small to keep.) Then all players count the cards in their fish basket pile.

Whoever caught the most fish wins the game.

Variations

HOUSE RULE: At our house, if you get the card you asked for, either from the other player or from the fishing pond, you get a free turn and may ask any player for another card.

FISH FOR FOUR: Do your children enjoy an element of risk? Instead of laying down pairs, players must collect all four cards in a set. This rule allows players to "steal" what another player asked for in an earlier turn.

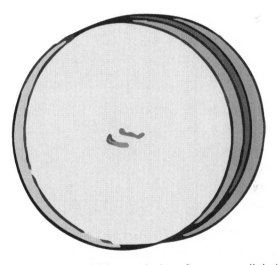

A card holder can help children with short fingers see all their cards.
Staple two plastic lids together, and decorate with stickers.
Slip cards between the lids to fan them out for easy viewing.

Concentration

MATH CONCEPTS: multiplication or fraction models, visual/spatial memory.

PLAYERS: any number.

EQUIPMENT: one deck of math model cards.

How to Play

Shuffle the cards and lay them all face down on the table, spread out in a single layer. The cards may be placed in an array or arranged in a haphazard cloud, as long as no card covers any other card.

On your turn, flip two cards face up. If the cards match, representing the same product or fraction, then you get to take the pair. If they do not match, leave the cards showing for a few seconds so all players can see what they are. Then turn them face down and let the next player take a turn.

Keep the cards you capture in a personal score pile. When all the cards are claimed, whichever player has collected the most is the winner.

Variations

HOUSE RULE: How will you handle the frustrating cycle where a player turns up new cards and sees that one of them would match a previously exposed card, but the other player grabs that pair, leaving the first player to try unknown cards again next turn? At our house, if you find a pair, you get a free turn and can flip over two more cards—which means every player exposes new cards that the next player can use. Free turns expire when there are ten or fewer cards left on the table, to keep one lucky player from claiming all the last pairs.

MIXED GROUPS: When playing with a wide range of ages, let the younger players flip three cards per turn and keep any two that match.

EQUIVALENT FRACTIONS: Instead of matching the fraction and pictures exactly, players may take any two cards that name the same amount of stuff. A card labeled ⅜ can match with a picture of ²⁄₄, since both of them are worth half of one whole thing.

History

Concentration is my favorite ice-breaker game for math club meetings because the game is quick to learn and easy to play in large groups. It is also a game that older children and adults can enjoy as much as the beginning students do. More than once, when my teenage daughter walked through the room where the younger children were playing, she asked to join in the game.

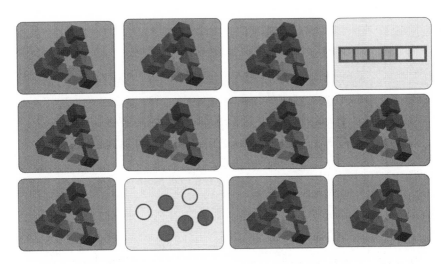

Turn up two cards. If they represent the same fraction or
multiplication expression, you may claim the pair.

War

MATH CONCEPTS: math models, less than or greater than.
PLAYERS: two or more.
EQUIPMENT: one or more decks of math model cards.

How to Play

Shuffle the math model cards and share them as evenly as possible among all players, or provide one deck per player. Players stack their cards face down on the table or floor. Then everyone flips their top card face up. The player with the greatest number wins the skirmish, capturing all the cards showing. Each player keeps a pile of prisoner cards.

If there is a tie for greatest card, all the players battle:

♦ Each player lays three cards face down, then turns a new card face up.

♦ The greatest of these new cards will capture everything played in that turn, including the face-down cards.

Because all players join in, someone who had a low card in the initial skirmish may win the battle. If there is no greatest card this time, repeat the three-down-one-up battle pattern until someone breaks the tie. The player who wins the battle captures everything.

Then the players go on to the next skirmish, again turning up the top card from each deck.

When the players have fought their way through the entire deck, count the prisoners (or compare the height of the stacks). Whoever has captured the most cards wins the game. Or shuffle the prisoner piles and play on until one player captures all the cards, or until all the other players concede.

Variation

You can also play Multiplication War or Fraction War with regular playing cards. See instructions in the next chapter (page 48).

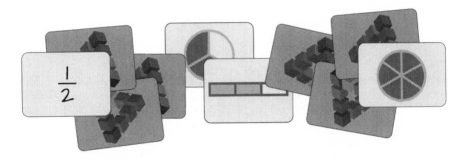

If the first cards played have the same value,
each player lays three cards face down
and then turns up a new card.

Make One

Math Concepts: fraction models, adding fractions.

Players: two or more.

Equipment: one deck of fraction model cards (use a double deck for more than four players), pencil and paper for keeping score.

Set-Up

Remove all the cards showing fractions equivalent to zero or one from your deck. If you're using my fraction card deck, remove the books for $\%_2$, $\frac{2}{2}$, $\frac{3}{3}$, $\frac{4}{4}$, and $\%$. Shuffle the remaining cards.

Place one card face up in the middle of the table. This card is the *last match*, and the player who claims it gets a double score.

How to Play

Deal all the cards out among the players as evenly as possible. It doesn't matter if some players have one more card than the others.

Players sort the cards in their own hands, looking for pairs of cards that add up to one whole thing. Collect your pairs face up on the table in front of you, so the other players can check your sums.

For instance, $\% + \frac{1}{3} = 1$, so those two cards would match. The cards don't have to have the same pictures—you can pair ½ of a pizza with ¾ of the cookies, or any other combination that adds up to one. In real world problems, addition only makes sense when you add pieces of the same type of whole thing, but for this game we're playing "fantasy fractions."

When all players are done matching, fan your remaining cards and hold them up so the other players can't see what you have. Play begins with the player on the dealer's left and continues clockwise around the table.

On your turn, draw a single card from the hand of the player on

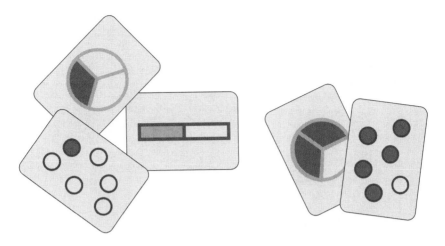

Always collect fractions in pairs. Don't match a
set like ½ + ⅓ + ⅙ = 1 because that will leave the
bigger fractions like ⅔ or ⅚ without partners.

your right. If the new card can pair up with a fraction in your hand to
make one, lay the pair with your score card collection. Otherwise, add
the new card to your hand. Then turn so the player on your left can
draw one of your cards.

If you get rid of all your cards, sit out the rest of the game until
it's time to record scores. Eventually all the cards will pair up except
one. The player holding this card claims the last match from the table,
ending that hand.

Scoring

At the end of a hand, count each player's score as follows:

- Each card in your collection is worth 1 point.
- The player who claimed the last match scores double, 2 points
  for each card collected.

Collect and shuffle the cards so the next player can deal a new hand.
Continue playing until someone reaches (or passes) 100 points to win

the game. If two or more players reach 100 points in the same hand, whoever has the highest score wins.

Variations

You can also play games like Go Fish or Concentration by matching pairs that total one. Or play Twelve Cards (page 28)—but because the Make One deck is smaller than the regular fraction models deck, use the Nine Cards variation.

History

Make One is based on the traditional childhood game of Old Maid (or Black Peter). If your kids enjoy online games, they can also practice adding fractions to total one with the interactive "Number Bond Fractions" game at Colleen King's Math Playground site.[†]

---

† *mathplayground.com/number_bonds_fractions.html*

Rummy

MATH CONCEPTS: multiplication or fraction models.

PLAYERS: two or more.

EQUIPMENT: one deck of math model cards, or a double deck for four or more players.

How to Play

Deal seven cards per player, and place the remaining cards face down as a draw pile. Turn up the top card of the deck to start the discard pile.

On your turn, you may either draw the top card from the deck or pick up the discard pile as far back as desired. But if you pick up more than the top discard, you have to meld the farthest-back card you take.

After drawing, you may *meld*—that is, place three or more matching cards face up on the table in front of you. If you have the fourth card in a set that has already been played (by any player), you may also lay that down in front of you.

Finally, put one card on the discard pile to end your turn. But if you discard a card that could have been played, any other player can call "Rummy!" and meld your discard.

If the deck runs out, take all the discards except the top one. Shuffle these cards and place them face down so the next player has a stack to draw from. Play continues until one player runs out of cards (either by laying them all down or by discarding the last one).

Scoring

Count each player's score as follows:

- ♦ Each card played on the table is worth 5 points.
- ♦ For every card remaining in the hand, subtract 2 points.
- ♦ The player who went out gets a bonus of 15 points.

You may play a single hand, just for fun. Or play several hands, and the first player to reach 300 points wins the game. Or set a different point goal based on how long you want the game to last.

Variation

FLOATER: You must have a discard to end the game, and this must be a card that could not be melded. If you lay down all your cards without a discard, you become a *floater*. Continue to play your turn—drawing or picking from the discard pile and melding cards—until you can go out with a proper discard. (But it is illegal to pick up just the top discard and immediately discard it again.)

History

According to the Pagat website, Rummy-style games first appeared in the early twentieth century. Like many card games, Rummy picked up different flavors as it traveled from one player to another over the years. Feel free to modify the rules above to fit your family's favorite way to play.[†]

---

† *pagat.com/rummy/rummy.html*

Conquer the Times Tables

The most effective and powerful way I've found to commit math facts to memory is to try to understand why they're true in as many ways as possible. It's a very slow process, but the fact becomes permanently lodged, and I usually learn a lot of surrounding information as well that helps me use it more effectively.

Actually, a close friend of mine describes this same experience: he couldn't learn his times tables in elementary school and used to think he was dumb. Meanwhile, he was forced to rely on actually thinking about number relationships and properties of operations in order to do his schoolwork. (E.g. I can't remember 9 × 5, but I know 8 × 5 is half of 8 × 10, which is 80, so 8 × 5 must be 40, and 9 × 5 is one more 5, so 45. This is how he got through school.)

Later, he figured out that all this hard work had actually given him a leg up because he understood numbers better than other folks. He majored in math in college and is now a cancer researcher who deals with a lot of statistics.

—BEN BLUM-SMITH

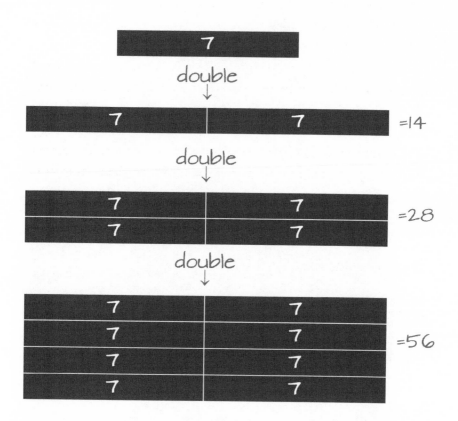

If you can double numbers, then you can
multiply anything times eight.

Quick Tip: Mental Calculation, Part One

MATH FACTS ARE THE BASIC relationships between one-digit numbers, such as 3 × 6 = 18 and 5 × 8 = 40. Many parents stress out over teaching the math facts, but children do not have to memorize long lists of number facts to be great at math. There is only one thing our children absolutely must learn: how to use whatever number facts they do remember to figure out the answers they forget.

For multiplication and division, if children learn the doubles and the square numbers, they can reason out everything else from those. (A *square number* is the answer when you multiply a number times itself, often represented by the columns and rows in an actual square of blocks: two rows of two, three rows of three, etc.) Once kids master the doubles and squares, show them how to split the difficult numbers up into easier ones.

And keep in mind these mental calculation tips…

The Doubles Family: ×2, ×4, and ×8

When we double a number, we get twice as much as we had at first. If we double it again, that will give us four times our original number. If we double it a third time, then we get double-four (which is eight) times as much as the number we started with.

Times two is double, times four is double-double, and times eight is double-double-double.

For example:

$$8 \times 6 = \text{double-double-double } 6$$
$$= \text{double-double } 12$$
$$= \text{double } 24$$
$$= 48$$

The Tens Family: ×10 and ×5

Please don't tell children to "just add a zero" when multiplying by ten. Instead, tell them to think of money: a penny times ten turns into a dime, and a dime times ten becomes a dollar. Any number times ten shifts to the next higher place value, so 5.37 × 10 = 53.7—we only add a zero if we need it to fill in an empty ones column.

Multiplying by ten is easy. And since five is half of ten, we can do any times-five calculation by doing half of times-ten.

For example:

$$5 \times 8 = \text{half of } 80$$
$$7 \times 5 = \text{half of } 70$$
$$5 \times 12 = \text{half of } 120$$
$$2.4 \times 5 = \text{half of } 24$$

[To be continued in Chapter 4.]

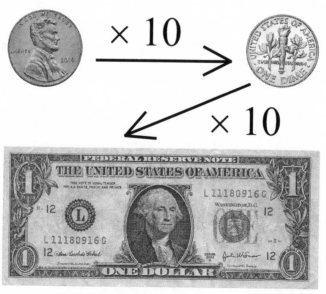

Multiplying by ten transforms a number
into the next greater place value.

Galactic Conquest

MATH CONCEPTS: multiplication math facts, rectangular area.
PLAYERS: two to four.
EQUIPMENT: game board or graph paper (1 cm squares for two players, or ¼ inch squares for more), two six-sided dice, colored markers.

How to Play

Each player will need a colored marker to shade in the game board squares, and the colors must be different enough to be easily distinguished. One sheet of graph paper represents the galaxy. Players each color a large dot on one corner of the grid, as far apart from each other as possible, to represent their home planets.

On your turn, roll the dice. Using those numbers as length and width, draw a rectangle that shares at least one corner with your current territory. Your new rectangle may not overlap squares already claimed by any player. Inside the rectangle, write the area (length × width) of your newly conquered space.

The game ends when a player cannot draw a rectangle to match the dice. Players add up the areas of all their rectangles, and whoever has conquered the most territory wins.

Variations

FOR BEGINNERS: You can play with a partial deck. Remove the cards for numbers they haven't studied.

HOW CLOSE TO 100?: (A cooperative game.) Mark a 10 × 10 square on your graph paper. Players take turns rolling the dice and coloring in a rectangle with the dimensions on their dice. Pack the rectangles as tightly as you can. How close can you get to coloring the full 100 squares before you roll a product that won't fit?

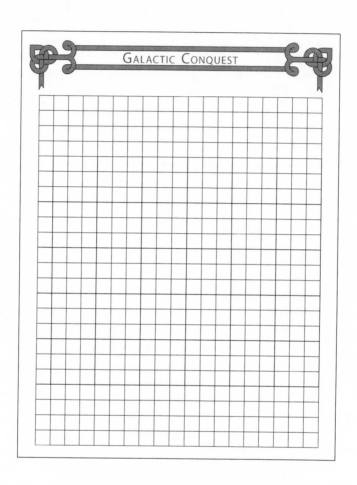

The Galactic Conquest game board from the free
Multiplication & Fraction Printables download file.[†]

---

†   *tabletopacademy.net/free-printables*

GALACTIC BLOBS: Players may draw any single, connected shape that covers the area representing the product of the rolled dice. Every square of the shape must share at least one side with some other part of the shape. For instance, a player could draw the L-shaped area of a chess knight's move, but not the diagonal squares of a bishop's path because those only meet at the corners.

WARP SPEED: To practice the hardest multiplication facts, remove the aces, twos, threes, and tens from a deck of math cards. Shuffle the rest of the deck and place it face down. Draw two cards on each turn. Mark your rectangle, and then lay the discards face up. Play until the deck runs out or until two players are forced to pass in consecutive turns.

History

Dice, graph paper, and the rectangular model of multiplication make a natural combination, and many teachers have shared versions of this game online. The "How Close to 100?" variation above comes from Jo Boaler's YouCubed website, a great resource for ways to teach math while encouraging a growth mindset.[†]

John Golden puts a different twist on area games with Area Battle, a War-style game for which students create their own cards.[‡]

[†] *youcubed.org/task/how-to-close-100*
[‡] *mathhombre.blogspot.com/2011/10/area-battle.html*

Multiplication War

MATH CONCEPTS: multiplication math facts, less than or greater than.
PLAYERS: two or more.
EQUIPMENT: math cards, one deck per player.

How to Play

Players shuffle their own decks and place them face down on the table or floor. (We play on the carpet.) Then everyone turns their top two cards face up and multiplies the numbers. The player with the greatest product wins the skirmish, capturing all the cards showing. Each player keeps a pile of prisoner cards.

If there is a tie for greatest product, push all the cards into the center. Then the players go on to the next skirmish, again turning up the top two cards from each of their decks. The winner of that skirmish takes the center cards. If there is no winner this time, repeat until someone wins a battle and captures the whole mess.

When the players have fought their way through the entire deck, count the prisoners (or compare the height of the stacks). Whoever has captured the most cards wins the game. Or shuffle the prisoner piles and play on until one player captures all the cards, or until all the other players concede.

Variations

Before play begins, the youngest player picks off part of the deck to reveal a hidden card. This determines whether the goal of the game will be high numbers or low.

If the exposed card is:

♦ 1–5: Lowest product wins each skirmish.

♦ 6–10: Highest product wins, as described above.

FRACTION WAR: Players turn up two cards and make a fraction, using the smaller card as the numerator. (This is the same as dividing one card by the other.) Greatest fraction wins the skirmish.

IMPROPER FRACTION WAR: Turn up two cards and make a fraction, using the larger card as the numerator.

MATH WAR TRUMPS: Players alternate choosing "trump" for the math card battles. After the cards are turned up, the player whose turn it is gets to say which operation (multiplication or fractions) to do.

ADVANCED MULTIPLICATION WAR: Turn up three (or four) cards for each skirmish and multiply them all together. Greatest product wins. Players may end up with one or two cards at the end of their decks. Draw enough cards from your prisoner pile to finish the last skirmish.

History

You can adapt this classic children's card game to practice a variety of number facts. A math card deck contains forty cards, so a single game of Multiplication War lets a child practice twenty problems. And if your children are like mine, they rarely want to stop after one time through the deck.

My students need extra practice on the hard-to-remember calculations. To give a greater challenge to older children, I make each player a double deck of math cards, but I remove the aces, twos, and tens. This gives each player a fifty-six-card deck full of the toughest problems. For even more of a challenge, use the face cards as additional numbers: jack = eleven, queen = twelve, king = thirteen.

X									

The *Multiplication & Fraction Printables* file
includes blank times tables for practicing math facts
up to 10 × 10 or 12 × 12. Since the edges are also
blank, players can scramble the rows and columns
or use larger numbers for added challenge.

Times-Tac-Toe

MATH CONCEPTS: multiplication math facts, times tables.

PLAYERS: best with two.

EQUIPMENT: printed blank times-table chart, deck of math cards, colored markers or a set of matching tokens for each player.

How to Play

Label the top row and left column of your blank times-table chart with the numbers 1–10, in numerical order or mixed around.

On your turn, flip two cards. Multiply them and find the corresponding square on the times table. If that square is blank, write in the product with your colored marker. Or say the product aloud while you cover that square with one of your tokens. If a player writes or says the wrong answer, the other player may challenge, give the correct product, and then take that square.

Sometimes you will have a choice of two squares, but you may mark only one of them. On the other hand, if there are no remaining spaces for your product, then you lose that turn. The first player to mark four squares touching (with no gaps) in a row—horizontal, vertical, or diagonal—wins the game.

Variations

For a longer game, play until someone marks five squares in a row.

If you prefer teaching the multiplication facts up to 12 × 12, you can include face cards in your deck: jack = 11, queen = 12, and king = wild card. A player who turns up a king may use any number in its place.

TIMES TABLE GOMOKU: No math cards. Players may choose any unclaimed square and write in the product of the row and column numbers. For a tougher challenge, one player fills the top row with any numbers greater than five, and the other player chooses numbers for the left-column squares. As in traditional Gomoku, it takes five in a row to win.

HUNDRED-TAC-TOE: Instead of a times-table chart, use a printed 100 chart. On your turn, flip one card, and you may mark any multiple of that number which has not already been taken. Say the factors of your multiple: if you draw a two, and you want to mark forty-six, say, "Two times twenty-three." Whenever you get four (or more) squares in a row, mark them with a solid line. Go through the whole deck, and the player who marks the most lines wins.

History

People all around the world have played make-a-row games since ancient times, with a wide assortment of rules. Claudia Zaslavsky surveys the history of such games in *Tic Tac Toe and Other Three-in-a-Row Games from Ancient Egypt to the Modern Computer.*

The Product Game

MATH CONCEPTS: multiplication math facts, factors, multiples.
PLAYERS: only two.
EQUIPMENT: printed game board, colored markers or a set of matching tokens for each player, two acrylic gemstones or other small tokens to mark the factors.

How to Play

The first player places a stone on any one of the factors at the bottom of the board. The second player places the other stone on a factor—the same or different—and then marks the product of those two numbers.

On each succeeding turn, a player moves just one stone to a new number and then marks the product of those two factors. If both players agree that all possible moves have already been colored in, the player whose turn it is may make a fresh start by moving both stones.

Whichever player marks four (or more) squares in a row—horizontal, vertical, or diagonal—wins the game. The squares must touch each other at edges or corners, with no gaps. If neither player can make four in a row, then the player who has the most sets of three in a row wins.

Variations

PATHWAYS: One player "owns" the top and bottom of the game board, while the other player claims the right and left sides. The first player who can mark a path of squares across the board (top to bottom for one, side to side for the other) wins the game. The pathway squares must all connect by sharing a side or corner.

PRODUCT GAME BOXES: On your turn, you may color one line segment on any single side of the square that contains the product of your numbers. When you draw a fourth line, completing a square, then you

1	2	3	4	5	6
7	8	9	10	12	14
15	16	18	20	21	24
25	27	28	30	32	35
36	40	42	45	48	49
54	56	63	64	72	81

1 2 3 4 5 6 7 8 9

The Product Game board, traditional style.
My printables file also offers a spiral version, which
moves the more difficult products into the valuable
central squares to encourage their use.

color that square with your color and take a free turn. The game ends when either player cannot color a line to match the numbers. Whoever has colored in the most boxes wins.

History

I first saw The Product Game on the 1987 *Square One TV* show, where the game was called "But Who's Multiplying?" You can watch contestants Brigitte and L'Wanda play the game in Season 1, Episode 8, on YouTube. Or your children can play the game online against the NCTM Illuminations computer.[†]

Marilyn Burns posted the Pathways game—which is a cross between The Product Game and the strategy game Hex—on her Math Blog, along with a lesson plan and several game boards to choose from.[‡]

The Boxes game is based on Joshua Greene's post "Dots and Boxes Variation" on Three J's Learning blog. And Greene posed several questions prompting students to examine and extend The Product Game in "Times Square Variations."[§]

† *youtube.com/watch?v=p7C2w2WAeVU*
illuminations.nctm.org/Activity.aspx?id=4213
‡ *marilynburnsmathblog.com/wordpress/the-game-of-pathways*
§ *3jlearneng.blogspot.com/2016/02/dots-and-boxes-variation.html*
3jlearneng.blogspot.com/2015/11/times-square-variations-math-games.html

1	2	3	2	4	6	3	6	9
4	5	6	8	10	12	12	15	18
7	8	9	14	16	18	21	24	27
4	8	12	5	10	15	6	12	18
16	20	24	20	25	30	24	30	36
28	32	36	35	40	45	42	48	54
7	14	21	8	16	24	9	18	27
28	35	42	32	40	48	36	45	54
49	56	63	56	64	72	63	72	81

1 2 3 4 5 6 7 8 9

Ultimate Multiple-Tac-Toe game board.

Ultimate Multiple-Tac-Toe

MATH CONCEPTS: multiplication math facts, times tables, thinking ahead.

PLAYERS: only two.

EQUIPMENT: printed or hand-drawn Multiple-Tac-Toe chart, colored markers, two acrylic gemstones or other small tokens to mark the factors.

Set-Up

If you are starting with a blank Multiple-Tac-Toe chart, give children time to write in the multiples in pencil. Each small Tic-Tac-Toe board contains the first nine multiples of one counting number, which are the answers in that number's times table. As they fill in these numbers, children internalize the structure of the game board, which makes playing the game go smoothly.

How to Play

The player who is marking X goes first, placing a stone on any one of the factors at the bottom of the board. The second player places the other stone on a factor—the same or different—and then marks an O on the product of those two numbers. On each succeeding turn, a player moves just one stone to a new number and then marks the product of those two factors.

Early in the game, players may choose from more than one square—for instance, for 3 × 7, the player may mark the times-three or times-seven board. As the game progresses, options will grow increasingly limited. If the only unmarked square is on a small Tic-Tac-Toe board that has already been won, the player must still mark there. But if both players agree that all possible moves have already been marked, the player whose turn it is may make a fresh start by moving both stones.

If you win one of the small Tic-Tac-Toe boards, mark over the whole thing with a large X or O. If a small board ends in a draw, mark it with a large C for "cat's game"—and that counts as a win for either player.

The first player to claim three large squares (that is, three of the small boards) in a row wins the game. With a cat's game acting as a wild card, it's possible that both players can make a row in the same turn: double-win!

Variations

FAST GAME: Players mark all the products of their factor numbers at once, wherever they appear on the board. So a player who has the factor stones on two and nine will mark every 18 on the whole game board.

ULTIMATE TIC-TAC-TOE: This game is a pure logic puzzle, without numbers. Use a blank Multiple-Tac-Toe board. The first player marks any square on any of the small Tic-Tac-Toe boards. On each succeed-

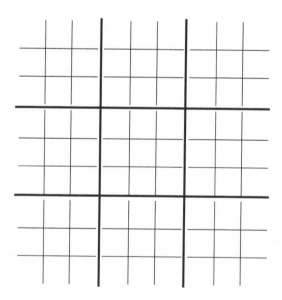

The Ultimate Tic-Tac-Toe game board has no numbers

ing turn, the player must mark a square on the small board that corresponds with the position just marked.

For example, if the first player marks the top right square on one of the small boards, then the second player may choose any unmarked square in the top right board. But if the top right board has already been won, then the second player may go to any square on any board.

History

Ben Orlin wrote about Ultimate Tic-Tac-Toe on his delightful Math with Bad Drawings blog. Federico Chialvo modified The Product Game to create the Fast Game version above, and then wrote a follow-up post analyzing how common factors and multiples affect the strategy of his game. And Orlin wrapped up the discussion with a post about the difference between a puzzle and a game.[†]

[†] *mathwithbaddrawings.com/2013/06/16/ultimate-tic-tac-toe*
artofmathstudio.wordpress.com/2013/08/30/multiplication-tic-tac-toe
artofmathstudio.wordpress.com/2013/10/31/revisiting-multiplication-tic-tac-toe-common-factors-and-multiples
mathwithbaddrawings.com/2013/11/18/tic-tac-toe-puzzles-and-the-difference-between-a-puzzle-and-a-game

Multiplication and division are *inverse operations,* which means they are two ways of looking at the same mathematical relationship. Can you find multiplication, division, and fractions in how the Sierpinski triangle changes from one stage to the next?

Mixed Operations

Most people like games, so that's an easy place to begin. At first the games can be the sweetness that helps the math medicine go down. Over time perhaps you can find the sweetness in the math itself—in a problem that inspires you to work and struggle until you finally get it, just for your own satisfaction.

Imagine a world in which we play around with math enough that a child recognizes her love of math, even if she's slow at learning new math topics in school. We would recognize that, although it comes more easily to some, we can all find our passion for its beauty, and work at learning it just for joy.

Remember that joy and passion lead to more learning than duty ever did.

—Sue VanHattum

Quick Tip: Mental Calculation, Part Two

IN CHAPTER 3, WE COVERED two mental multiplication families. All the remaining times tables are related to each other by one key principle: the distributive property. I explain the distributive property in Chapter 6 (page 113), so here I'll skip straight to the mental math tips:

The Distributive Family: ×3

Because 3 = 2 + 1, three of any number is the same as double that number plus one more of that number.

    3 × 7 = double-7 + one more 7

    15 × 3 = double-15 + one more 15

    3 × ½ = two halves + one more half

The Distributive Family: ×6 and ×7

Because 6 = 5 + 1, six times any number is five of that number plus one more of that number.

6 × 8 = five 8s + one more 8

6 × ⅕ = five fifths + one more fifth

(–8) × 6 = the negative of five 8s + one more (–8)

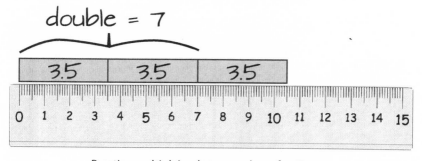

Practice multiplying large numbers, fractions,
decimals, and negative numbers, too.
For 3 × 3.5: double 3.5 is 7, and then add one more 3.5.

You can use other distributive patterns, too. For instance, 25s are easy when grouped in sets of four, so 7 × 25 = four 25s + three more 25s.

In the same way, 7 = 5 + 2, so any number times seven is five of the number plus two more copies.

$$7 \times 5 = \text{five 5s + two more 5s}$$
$$50 \times 7 = \text{five 50s + two more 50s}$$
$$7 \times 1.2 = 5 \times 1.2 + 2 \times 1.2$$

The Distributive Family: ×9, ×11, and ×12

Multiplying by ten is easy. Even though our last three times tables involve bigger numbers, they are related to times-ten. Which makes them relatively easy, right?

$$9 = 10 - 1$$
$$11 = 10 + 1$$
$$12 = 10 + 2$$

So you know what to do:

$$9 \times 7 = 10 \times 7 \text{ take away one 7}$$
$$11 \times 2.3 = 10 \times 2.3 \text{ and one more 2.3}$$
$$12 \times 60 = \text{ten 60s + two more 60s}$$

Practice Hard to Make Doing It Easy

Take the time to talk through these patterns and work many, many, many oral math problems together with your children. Discuss the different ways you can find each answer. Notice how the number patterns connect to one another.

When you are practicing each family of rules, be sure to challenge each other beyond the normal times-table facts. Try multiplying two- or three-digit numbers, fractions, decimals, or negative numbers. Make it into a game: students love the chance to stump their parents. Try working mental multiplication puzzles while you are doing dishes, or on the way to soccer practice, or whenever you can find a spare moment.

For more on this conversational approach to multiplication skills, read the "How to Conquer the Times Tables" series on my Let's Play Math blog.†

† denisegaskins.com/tag/times-table-series

Salute

MATH CONCEPTS: multiplication, division, factors, inverse operations, speed of recall.
PLAYERS: three.
EQUIPMENT: one deck of math cards.

How to Play

This game reinforces the inverse relationship between multiplication and division: If you know the product of two numbers, and you know one of the numbers, you can divide to find the other number. Inverse operations such as multiplication and division are closely related and should be studied together.

Players sit where they can all see each other's faces. Dealer gives one card, face down, to each of the other two players. Players *must not look at their own cards.*

Then the dealer says, "1, 2, 3, Salute!" On the word "Salute," players raise their cards to their forehead, with the numbers facing outward so the other players can see. The dealer multiplies the two cards and says, "The product is __."

The first player to (correctly) say the number on his or her own card wins the trick and collects both cards. Keep your captured cards face up beside you, to avoid confusing them with future hands. Then the dealer passes the deck to the left, and that player deals the next pair of cards.

When the deck is finished, the player who has collected the most cards wins the game.

Variations

For beginners, play with a partial deck: remove the cards for numbers they haven't studied. For older students, play with the face cards for a greater challenge: jack = eleven, queen = twelve, and king = thirteen.

TRIPLE SALUTE: Deal a card to all three players. When the dealer says "Salute," all players raise their cards. Multiply the two cards you can see and say the product out loud. The first player to deduce his or her own card's value from these clues and name it aloud takes the trick and collects all three cards.

HOUSE RULE: Do your kids get frustrated by math games that require speed? Have the dealer supervise two hands before passing the deck. With the first pair of cards, the player to the dealer's left tries to name his or her number without having to race the other player. For the second pair, the other player gets to guess. A correct answer entitles the guesser to both cards from that hand. If the answer is wrong, the dealer shuffles those cards back into the deck.

History

Constance Kamii wrote about this game in *Teaching Children Mathematics* magazine.[†]

Usually I avoid any math game that depends on speed, but Salute gives students a wonderful introduction to thinking about inverse operations. Division is the inverse of multiplication, so children can solve division problems by thinking, "How many would I have to multiply that number by, to get this product?"

---

† *highbeam.com/doc/1G1-110906838.html*

Distributive Dice

MATH CONCEPTS: addition, subtraction, multiplication, distributive property, rectangular area, multistep mental math.

PLAYERS: two to four.

EQUIPMENT: graph paper (1 cm squares for two players, or ¼ inch squares for more) or the Galactic Conquest game board, three six-sided dice, colored markers.

How to Play

Each player will need a colored marker to shade in the game board squares, and the colors must be different enough to be easily distinguished. As in the Galactic Conquest game (page 45), players share a single sheet of graph paper. Each player colors a large starting dot on one corner of the grid, as far apart from the others as possible.

On your turn, roll all three dice. Choose two of the numbers to add or subtract, and that answer will form one side of your rectangle. The third die gives the other side. Draw the rectangle on the graph paper so that it shares at least one corner with your current territory and does not overlap squares already claimed by any player. Inside the rectangle, write the area of your newly conquered space.

The game ends when a player cannot draw any rectangle to match the dice. Players add up the areas of all their rectangles, and whoever has conquered the most territory wins.

Variations

DAMULT DICE: (Any number of players.) No graph paper needed. On your turn roll three dice: choose two to add, multiply by the third, and then add that many points to your score. If you have enough dice and enough space to keep them separate, players can all roll at the same time. The first player to reach 300 points wins the game—or if two

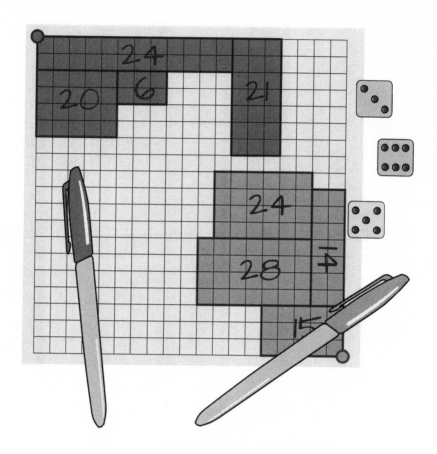

Distributive Dice game in progress. Dark Red's turn,
and the dice have been rolling low—but this time,
Red has two big numbers to play with.
Would you move to gain the most territory, or
try to block Purple's ability to grow?

players pass 300 points in the same turn, then the highest score wins.

DOUBLE DAMULT: Roll six dice. Use any combination of addition, subtraction, multiplication, and division to calculate your score—but you have to use each operation at least once.

MASQUERADE: (For two players.) Roll three dice but keep the numbers hidden. Add two, multiply by the third, and then tell your opponent the score. The other player gets three guesses to try to win the number of points on your dice. For each correct guess, reveal that die, and the other player scores that many points. If two dice are the same, only show one per guess. Play five rounds, and the highest score wins.

History

Dan Finkel invented the Damult Dice games and shared them on his Math for Love blog.[†]

John Golden modified the game, and I tweaked that to make my Distributive Dice version. Golden played the game with a fifth-grade class, and the students created several more variations, including Masquerade. Read more of their ideas in "Multiplying Game Possibilities" at Math Hombre blog.[‡]

[†]   *mathforlove.com/2010/10/a-game-to-end-all-times-tables-drills-damult-dice*

[‡]   *mathhombre.blogspot.com/2012/04/multiplying-game-possbilities.html*

91	92	93	94	95	96	97	98	99	100
81	82	83	84	85	86	87	88	89	90
71	72	73	74	75	76	77	78	79	80
61	62	63	64	65	66	67	68	69	70
51	52	53	54	55	56	57	58	59	60
41	42	43	44	45	46	47	48	49	50
31	32	33	34	35	36	37	38	39	40
21	22	23	24	25	26	27	28	29	30
11	12	13	14	15	16	17	18	19	20
1	2	3	4	5	6	7	8	9	10

Many children find the bottoms-up hundred chart more
logical than the traditional top-down version.
The free *Multiplication & Fraction Printables* file contains both styles.[†]

[†] *tabletopacademy.net/free-printables*

Leapfrog

MATH CONCEPTS: addition, subtraction, multiplication, division, multistep mental math.
PLAYERS: any number.
EQUIPMENT: printed hundred chart, one regular deck of playing cards, one small toy or token per player.

How to Play

Deal four cards to each player, and place all the tokens near the first (single-digits) row of the hundred chart. Turn the rest of the deck face down as a draw pile.

On your turn, draw one card. Then choose your move:

♦ Use any three of the number cards in your hand to form a two-step calculation that equals any number in the next-higher row of the chart. (Face cards have no number value in this game.) Show your cards and say your equation, then jump your token to that square. Discard one of the cards you didn't use.

♦ Play a face card by laying it on the table in front of another player, who must move his or her token down to any space in the next-lower row. The face card counts as your discard, ending your turn.

♦ If you can't make a jump or play a face card, choose any card from your hand to discard.

Two or more players' tokens may share a square on the game board. But if your token gets moved down by a face card, you may not use the same calculation twice in succession. You may jump back up to the same square you used before only if you find a different way to calculate that number.

If the deck runs low, shuffle the discards back into the draw pile. The first player to leap from row to row all the way to the top and then make a number greater than one hundred jumps off the chart and wins the game.

Variations

HOUSE RULE: Do your children find the higher rows too hard to reach? Allow players to use four cards in calculating jumps to the top four rows (numbers past 60).

BONUS JUMPS: After you move, if you can use the same three cards to calculate a number in the next row, you can jump again.

SUPERFROG: You can play face cards on other players (moving them back one row) or on yourself (without moving down). Any player who collects six face cards gains one-time superpowers. When you lay down the sixth face card, you may jump to the highest number you can calculate using all four cards remaining in your hand. If you can make a number greater than one hundred, you win the game. If you can't jump off the board, then discard the face cards and continue normal play from your new position.

LEAPFROG SOLITAIRE: Using a math card deck (without face cards), turn up six cards. Choose any three cards to make a two-step calculation, as above, but don't discard. Keep re-using the same six cards. Can you jump all the way up the chart?

History

My version of Leapfrog is adapted from fourth-grade teacher Chris Brewer's Leap Frog Cover-up board game, which Alice Wakefield shared in *Early Childhood Number Games: Teachers Reinvent Math Instruction.*

Averages

MATH CONCEPTS: addition, division, mean, median, mode.

PLAYERS: two or more.

EQUIPMENT: one deck of math cards, pencil and paper for keeping score.

How to Play

Deal seven cards to each player. Consider these cards as your data, and choose the type of average that will give you the highest score:

- ◆ Mean = the number each data point would be, if the total were shared out equally. Add all the numbers, divide that sum by seven, and then round to the nearest whole number.

- ◆ Median = the middle number when the cards are arranged from least to greatest.

- ◆ Mode = the number that appears most often. Not all sets have a mode.

Tell which type of average you are using, and add that number of points to your score. After each player has scored, pass the deck to the next dealer. When every player has had at least one chance to deal, whoever has the highest score wins the game. Or play until someone reaches 50 points (or some other agreed-upon target).

Variation

Or award 1 point to the player with the largest of each average. The first player to reach 15 points wins.

AVERAGE TRUMPS: After looking at his or her cards, the player to the dealer's left chooses which type of average to score for that hand.

History

In statistics, we have three ways of looking at the average of a set of data, each of which conveys different information about relationships among the numbers. One of the easiest ways to "lie with statistics" is to be careless of which average you are talking about.[†]

This game is adapted from the M&M&M's activity in *Acing Math (One Deck at a Time)* by The Positive Engagement Project.[‡]

---

† *archive.org/details/HowToLieWithStatistics*
‡ *pepnonprofit.org/mathematics.html*

Twenty-Four

MATH CONCEPTS: addition, subtraction, multiplication, division, order of operations, multistep mental math.

PLAYERS: any number.

EQUIPMENT: one deck of math cards.

How to Play

Deal four cards to each player, face down. The players must leave the cards face down until everyone is ready. Set the rest of the deck to one side.

At the dealer's signal, all players pick up their hands and look at the cards. Each player tries to combine all four numbers on the cards to make twenty-four (or another designated target number).

Players may add, subtract, multiply, or divide the numbers in any order, but they may not put two cards together to make a two-digit number. Each card may be used only once in the calculation.

For example, a hand of 4, 3, 7, and 9 could make:

$$(9 \times 3) - 7 + 4 = 24$$
or
$$(9 - 7) \times 3 \times 4 = 24$$
but not
$$(9 - 3) \times 4 = 24,$$
which ignores the 7 card.

This game has an element of luck. Some hands will not make twenty-four no matter how you combine the numbers. If all players seem stumped, the dealer should give each player one more card. The players may use all five cards in their hands or choose any combination of four. On rare occasions, the dealer may have to deal a second round of extra cards before any player can hit the target.

When you figure out a way to make twenty-four, lay your cards face up on the table. Explain your calculation so the other players can check

How might you combine these cards to make
the target number twenty-four?

it. The first player to make twenty-four using at least four cards in legal arithmetic calculations is the winner of that hand and gets to deal the next round. Or play several hands, scoring 1 point per hand, and the first player to score 6 points wins.

Variations

HOUSE RULE: Do your kids get frustrated by too many hands that can't make the target? Allow the ace to stand for either one or eleven, at the player's discretion. Or use the face cards as additional numbers: jack = eleven, queen = twelve, king = thirteen.

SLAP TWENTY-FOUR: To eliminate the element of chance, deal four cards face up in the middle of the table. All players use these cards, and whoever is the first to calculate twenty-four slaps the table. The player then explains the calculation and, if it is correct, wins the hand and scores a point. If there is no solution possible, then the first player to say "No solution" wins the point—but if another player then gives a solution, the first player gets nothing, and the second one gains 2 points.

TWENTY-FOUR WITH VARIABLES: For a faster-paced game, include the face cards as variables (wild cards). A face card in your hand may take any value that a number card might have, from one to ten.

Score Twenty-Four: Give each player a piece of paper and a pencil or pen, and deal eight cards face up on the table. Set the timer for 10 or 15 minutes. Each player writes on a piece of paper as many ways as possible to combine any of two or more of these numbers to make twenty-four. Each card may be used only once in each calculation. Every valid expression scores 1 point per card used.

Target Number in the Car: Each player needs a clipboard (or other hard surface to write on) with paper and pencil. Players take turns naming a number between one and twenty, until there are eight numbers named. Numbers may be repeated. All players write these game numbers on their paper.

Then the driver names a target number between one and one hundred. The driver also sets a time limit for the game—perhaps until the next gas station or rest area. Players try to make the target number as many different ways as they can. Each game number may be used only once in a calculation. Every valid expression scores 1 point per number used.

History

The Pagat website traces the card game Twenty-Four to Shanghai in the 1960s, so the game may have originated in China.

In 1988, Robert Sun created the commercial 24 Game, which uses special playing cards. The game comes in a wide range of levels, allowing students to practice topics from simple addition to fractions, decimals, and algebra. You can see sample games, hints, teaching tips, and tournament information online.

Of course, there is nothing magical about the number twenty-four—you may choose any number to be your target. But for beginners, twenty-four is a good number because it has an abundance of factors: 1×24, 2×12, 3×8, and 4×6. This means players potentially have many ways to reach the goal.

1	2	3	~~4~~	5	6	7	8
9	10	~~11~~	12	13	~~14~~	15	16
17	18	19	20	~~21~~	22	23	~~24~~
25	26	27	28	~~29~~	30	31	32
33	34	~~35~~	36	37	38	39	40
41	42	44	45	48	50	54	
60	64	66	72	75	80	90	96
100	108	120	125	144	150	180	216

Contig board game. My printables file also offers a spiral version, which moves the larger numbers into the valuable central squares to encourage more advanced mental math.

Contig

Math Concepts: addition, subtraction, multiplication, division, order of operations, multistep mental math.

Players: two or more.

Equipment: your choice of printable game board, three six-sided dice, pencil or marker(s), paper for keeping score.

How to Play

On your turn, roll all three dice. Use the three numbers and the basic arithmetic operations (+, −, ×, ÷) to form a two-step calculation that equals the number in any unmarked square on the game board.

Think of as many possible combinations as you can, and choose the highest-scoring square. Mark your answer on the game board with a large X. Say out loud how you calculated the number.

You score 1 point for the square you marked, plus 1 point for each already-marked square *contiguous* to your number's square—that is, touching any side or corner. The maximum score for any turn is 9 points. If all the numbers you can make have already been marked, you score a zero—but if anyone else can find a valid calculation using your dice, that player may challenge you, mark the square, and "steal" those points.

When another player thinks you made an arithmetic mistake, that person may challenge your answer before the next player rolls the dice. If your answer was wrong, the challenger takes the points you would have won, and you score zero. But if your calculation is correct, you get one bonus point for having withstood the challenge.

Play until each player has had ten turns, or five turns each for more than three players.

Whoever has the highest total score wins the game.

Variations

The most common variation I have seen is not to score a point for the marked square. Just score 1 point for each contiguous square that was previously marked, which makes the maximum possible score per turn only 8 points. I prefer the scoring system above, which awards at least 1 point for any valid calculation.

CONTIG-TAC-TOE: Two players mark numbers with X and O, and the first player to get four squares in a row wins. Rows may be vertical, horizontal, or diagonal. For a longer game, try five in a row.

MULTIPLAYER EXTENDED GAME: Keep playing until almost all the numbers are marked. Any player who gets a zero three turns in a row drops out of the game. When the last player gets a third strike, the game is over. There is no bonus for the last player, other than the extra turn(s).

History

F. W. Broadbent published the original article about Contig in *The Arithmetic Teacher,* May 1972. For as long as I can remember, our local homeschool group has held a series of Contig practices every spring. Then we host a "school" tournament, and the top two players in each grade level proceed to a regional tournament against other public and private school teams.[†]

---

† *jstor.org/stable/41188047?seq=1#page_scan_tab_contents*
maconpiattroe.org/vimages/shared/vnews/stories/54c95745dbc67/Contig%20Rules.pdf

Fractions and Decimals

Some time ago, I was told of a girl who had to measure out some dangerous drug. The amount required was one-sixth of some unit, but she didn't have a measure for one-sixth. However, she had measures for one-half and one-quarter. She thought, "Six is two plus four. So I'll measure out one-half and one-quarter and the sum of these should be one-sixth."

The patient, I believe, survived.

The remarkable thing is that this girl would probably have understood more about arithmetic if she'd never learnt it at all. It's clear to a fairly young child that he'll get less cake if the cake is shared out among four children than if it's shared between two; he'll get even less if it's shared among six. He'd have no doubt which to choose, if he were offered one-sixth of a cake, or three-quarters.

The depressing thing about arithmetic, badly taught, is that it destroys a child's intellect, and to some extent, his integrity. Before they're taught arithmetic, children won't give their assent to utter nonsense; afterwards, they will. Instead of looking at things and thinking about them, they make wild guesses in the hope of pleasing a teacher or an examiner.

—W. W. Sawyer

Quick Tip: How to Compare Fractions

To PLAY MANY OF THE games in this chapter, our children must be able to tell which of two fractions is greater than the other. Which fraction names the bigger chunk of stuff?

In many cases, the comparison is simple. If one player turns up ½ and the other gets ¾, these familiar numbers are easy to judge. If one player draws ⅞ and the other has ⅔, they can agree that seven pieces is more than two.

Unfortunately, other comparisons are difficult enough to turn the game into an argument. For example:

◆ Can your children figure out whether ¾ is greater or less than ⁵⁄₇?

Benchmark Numbers

We can judge many fractions by asking, "Is it more than a half or less than a half?" Familiar numbers like one-half, one, and zero are called *benchmark numbers* because they make comparisons easy.

For instance, ⅞ is getting close to one whole thing, but ¹⁄₁₀ is a tiny piece, barely more than nothing. So ⅞ is the clear winner.

How about ³⁄₇ versus ⁵⁄₉? A quick sketch of any fraction model shows that ³⁄₇ is less than half because fewer pieces are colored than blank. The same reasoning proves ⁵⁄₉ is more than half. Therefore, ⁵⁄₉ is the greater fraction.

Or if the debate had been between ¾ and ⁶⁄₇, we could think about how close the fractions are to one whole thing. ¾ needs one more fourth to make one whole. ⁶⁄₇ needs one more seventh. Since one seventh is the smaller piece, we know ⁶⁄₇ names more stuff.

As handy as benchmark numbers are, they won't solve every fraction comparison. In the puzzle above, ¾ and ⁵⁄₇ are both bigger than a half and smaller than one whole thing. We still don't know which fraction is greater.

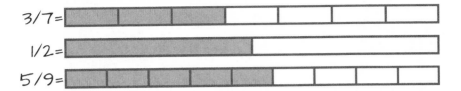

Children familiar with math models may be able to imagine the pictures. But when in doubt, take the time to draw bars.

Same-Sized Pieces

The top number (*numerator*) of a fraction tells us how many pieces we have, and the bottom number (*denominator*) tells us how small the pieces are—how many it takes to make one whole thing. When the pieces are different sizes, comparing fractions can be tricky. But when the denominators are the same, it's easy to see that the fraction with more pieces represents a greater amount of stuff.

So if we get stumped, one way we can compare two fractions is to make the denominators match. We need to turn at least one of our fractions into an *equivalent fraction*—a fraction with different numbers that still names the same quantity of stuff.

To make an equivalent fraction, we multiply the top and bottom numbers of the original fraction by the same amount. This is like cutting our pieces into smaller bits. We will have more pieces (a larger numerator) of smaller size (larger denominator), but as a group, they still represent the same chunk of the whole as our original fraction did. The number of pieces goes up in the same proportion as the size goes down, so both numerator and denominator must be multiplied by the same factor.

When two fractions share the same denominator, we say they "have it in common," and we call it their *common denominator*. The easiest way to find a common denominator is by multiplying the denominators of your two fractions.

slice
each
piece

2/3 ⟶ 6/9

When we cut a pizza into smaller slices, we get
more pieces. If we cut three times more pieces, each
slice will be one-third of the original size.

Which Is Greater: ¾ or 5⁄7?

To compare fourths and sevenths, we need to find a common
denominator, so we calculate $4 \times 7 = 28$. If we cut both fractions into
twenty-eighths, we will be able to compare them. We need to multiply
both parts of ¾ by 7, then multiply both parts of 5⁄7 by 4, and—finally—
see which of the fractions represents more twenty-eighths.

$$\tfrac{3}{4} = {}^{(3 \times 7)}/_{(4 \times 7)} = {}^{21}\!/_{28}$$
$$\tfrac{5}{7} = {}^{(5 \times 4)}/_{(7 \times 4)} = {}^{20}\!/_{28}$$
$${}^{21}\!/_{28} > {}^{20}\!/_{28}$$

Therefore, the player with ¾ wins this skirmish.

Nix the Trick: Cross-Multiplying

When we adults were in school, most of us learned to *cross-multiply*
fractions. By giving this technique a name of its own, our teachers
hoped to help us remember it. But they also made it harder for us
to see how it connected with what we had learned before. "Cross-
multiplying" is simply *making equivalent fractions.*

Just as in the example above, each fraction gets cut into smaller pieces in the proportion of the other fraction's denominator. The new fractions have more pieces of smaller size, and by using each other's denominator, we guarantee that the final fractions have pieces that match.

Since the denominators of our new fractions will be the same, we can ignore them, just as we could ignore the twenty-eighths above. We only have to compare the numerators to see which fraction wins.

Common sense can solve many fraction problems.
For instance, if two fractions have the same numerator,
then they have the same number of pieces.
So compare the denominators.
Which fraction is counting bigger pieces?

3/7 =

3/9 =

Domino Fraction War

MATH CONCEPTS: proper fractions, comparing fractions.
PLAYERS: two or more.
EQUIPMENT: one set of double-six or double-nine dominoes.

How to Play

Remove the double-blank tile. Turn all remaining domino tiles face down on the table and mix them around to make the wood pile.

Each player turns one tile face up and makes a proper fraction (smaller number on top) with the numbers on the two halves of the tile. The player who has the greatest fraction takes the other players' tiles prisoner, placing his own and all the captured tiles face up at his side.

If there is a tie for greatest fraction, turn that hand of domino tiles face down and shuffle them back into the wood pile. Then the players all turn up new tiles for the next skirmish.

When there are no longer enough dominoes on the table for every player to draw one, the players count their prisoners. Whoever has captured the most tiles wins the game.

Variations

Instead of requiring the smaller number to be on top, you can allow improper fractions, but be careful with blank sides. A blank is a zero, and a fraction may never have zero in the denominator, so the blank always has to go on top.

Domino tiles work for any Math War variation. Turn up a tile and add, subtract, or multiply the two numbers—whatever you want to practice.

DOMINO WAR TRUMPS: Beginning with the youngest and continuing around the table, players take turns saying "high" or "low" to tell

Is ⅔ greater or less than ⅘? Use a benchmark number: which fraction is closer to one whole thing? Which needs the smaller piece to make one?

whether the greatest or least fraction will win that round. Players draw tiles as described above but hide them from each other. After seeing his or her tile, the player whose turn it is declares the trump. Then all players display their tiles.

EXPONENT WAR: If you draw the 3|5 tile, will you make 3^5 or 5^3?

History (and a Puzzle)

Domino-like tile games seem to have originated in China, and they came to Europe through the great trading cities of Venice and Naples. Some game historians claim the European game was invented independently, because European domino sets are different from Chinese sets in several ways. (For instance, Chinese tiles come in suits, like a set of playing cards.) Dominoes spread across France and reached England in the late eighteenth century, where the game became a favorite pastime in British pubs.

Encourage your children to examine a set of domino tiles and describe what they notice. For example, every possible combination (double-0, 0|1, 0|2, etc.) is a single tile, but there are no duplicates: 0|1 is the same tile as 1|0.

Ask them, "If you bought a set of dominoes at a garage sale, how could you tell whether any of the tiles were missing? Can you figure out how many tiles there should be?"†

---

† *Spoiler: To find the answer, make a systematic list, and be careful not to count any of the combinations twice. A double-six set should have twenty-eight tiles, and a double-nine set will have fifty-five. A new set from the store may contain extra blank tiles, which can be decorated with paint or white nail polish to replace lost pieces.*

Fraction Train

MATH CONCEPTS: proper and improper fractions, comparing and ordering fractions.

PLAYERS: any number.

EQUIPMENT: one set of double-six or double-nine dominoes.

How to Play

Remove the double-blank tile. Turn all remaining domino tiles face down on the table and mix them around to make the wood pile. Each player draws five to seven tiles *but does not look at them.* Players arrange their tiles in a row (train) of fractions, as shown. When all players are ready, turn the tiles in your train face up without changing which side of the tiles is at the top of each fraction.

Your goal is to make your fraction train increase from left to right, but of course it will be mixed up to start with. On your turn, draw one tile and turn it to create a *proper fraction* (numerator less than denominator) or *improper fraction* (numerator equal to or greater than denominator) with the numbers on the two halves of the domino.

You have three choices:

♦ Use the new fraction to replace one of the domino tiles in your train. Then discard the old one, mixing it into the wood pile.

♦ If you don't want to use the new fraction, put it back and mix up the tiles.

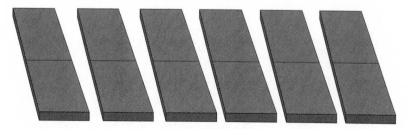

A domino number train, ready to flip and play.

♦ *Instead of drawing a new tile,* you may use your turn to rotate one of your current tiles, inverting the numerator and denominator of that fraction.

The first player to complete a train of fractions that increases from left to right wins the game.

Variations

HOUSE RULE: Decide how strict you will be about the "increases from left to right" rule and repeated numbers. Will you consider equivalent fractions as part of a valid train? Or will the player have to keep trying for a domino to replace one of the equivalents?

MATH MODELS NUMBER TRAIN: Play with either the multiplication or fraction deck of math model cards. With the multiplication cards, the value of the products must increase from left to right.

History

Fraction Train has been a favorite game in my middle school math club for years. We began by playing the game with Fraction Bars, but dominoes are cheaper and offer more variety.[†]

---

† *fractionbars.com*

Fraction Train with Cards

Math Concepts: proper and improper fractions, comparing and ordering fractions.

Players: any number.

Equipment: one deck of math cards for two players, or a double deck for more players or longer trains.

Set-Up

This game is best played on the floor, so everyone has plenty of room. Turn all the cards face down and mix them around in the middle of the playing area to create a fishing pond.

Each player draws out six (or eight, or ten) cards, but *does not look at them*. Lay the cards face down in the shape of a series of fractions, one card for each numerator or denominator, like this:

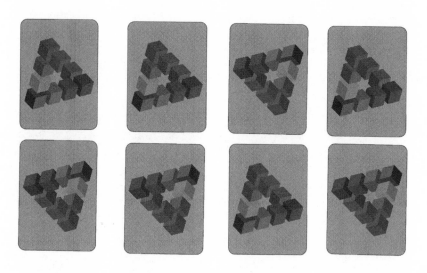

If you draw eight cards, arrange them to make a
row of four fractions, ready to flip and play.

How to Play

All players turn their cards face up at once, leaving them in their original positions. Your goal is to make your fraction train increase from left to right.

On your turn, draw one card from the fishing pond. You may use this card to replace any of the cards in your fraction train, the numerator or denominator of one of your fractions. Place your old card face down and mix it back into the pond.

If you do not wish to use the card you drew, mix it back into the pond, but you may not draw another card until your next turn.

The first player to complete a train of fractions that increases from left to right wins the game.

Variations

To play in a limited space, players should draw their fraction trains on paper or a whiteboard. Put an empty rectangle for each numerator and denominator. You only need one deck of cards, no matter how many people play. On your turn, choose one card and write that number into one of your rectangles—either in an empty one, or replacing a number you had previously written—and then mix the card back into the fishing pond.

HOUSE RULE: Decide how strict you will be about the "increases from left to right" rule and repeated numbers. Will you consider equivalent fractions as part of a valid train? Or will the player have to keep trying for new cards to replace one of the equivalents?

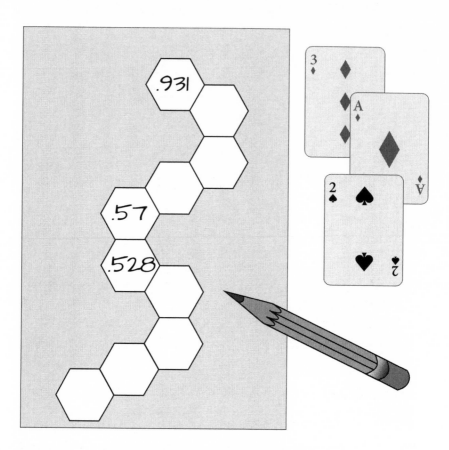

You can make .123 or .231 or .312, and so on.
Which number would you choose?

Decimal Point Pickle

Math Concepts: decimals, place value, numerical order.
Players: any number.
Equipment: one deck of math cards, whiteboard and markers for each player, or pencils and blank paper.

Set-Up

Each player draws a path with ten spaces big enough to write in three-digit numbers. Decimal Pickle paths can be utilitarian or creative: a simple row of boxes, a curvy chain of circles, a series of stair steps, a caterpillar of ovals with legs, or a string of flowers with open centers for writing in. But every path needs to have ten spaces with a clear beginning and end.

If you draw the Pickle paths on paper, you can laminate these drawings or slip them into sheet protectors for repeated play. But if you make a new drawing each time, then the game can express the children's personalities as their artistic skills develop.

How to Play

Remove the tens from the deck, but add the queens to represent zeros (or leave in the tens and count them as zeros). Shuffle the deck and place it face down where all players can reach.

On your turn, flip over the top card. If the card is black, stop flipping: that is your number. If the card is red, keep it and flip up an additional card. Again, if it's black, you stop, but if it's red, flip over one last card. You never flip more than three cards.

Now, arrange your card(s) to make a decimal number less than one. If you've drawn a single (black) card, you have no choice—for instance, if you draw a black seven, you must make .7. If you draw a red card and then a black card, you have two options—for instance, a red two and a black six lets you make either .26 or .62. Queens are zeros, so the smallest number you can make is a queen = .0, and the largest is three nines = .999.

Say your number, and write it into one of the spaces on your path. Always make sure the numbers increase from the beginning of your path to the end. If there's no place to fit in your number and keep the least-to-greatest pattern, then you miss that turn.

Discard face up next to the draw pile. If you run out of cards in the draw pile, shuffle the discards to replenish it. The first person to fill a path, with all the numbers in order from least to greatest, wins the game.

Variations

Play that you always flip over two or three cards (color doesn't matter). Or play without the three-card limit, so you might hit a 10-digit or longer decimal.

Or leave the 10s in the deck, letting them take two decimal digit places—so if you turn up a 7 and a 10, you could make the numbers .710 or .107. If you draw more than one 10 card, you may end up with a four-digit (or longer) decimal.

HOUSE RULE: Decide how strict you will be about the "increasing order" rule and repeated numbers. Can a player use both .6 and .60 as part of a valid train? Or will the player have to keep trying for new cards to replace one of the equivalents?

History

Blogger John Golden invented this version of the Number Train game. He writes:

The fifth-grade came up with two names, Destination Elimination (which I like because it rhymes), and Decimal Pickle. This was suggested by a student whose answer for everything is "pickle." But here, it reminded me of a childhood game that none of the kids knew but kind of fits: the baseball game Pickle.

As often with a new game, I played me vs. the class first. It was clear that the blackjack-esque possibility of extra cards was exciting, and they quickly got the idea that it was a big advantage. I didn't castigate anyone for saying "point two three" but often asked, "So how do you say that number?"[†]

---

† *mathhombre.blogspot.com/2010/05/decimal-point-pickle.html*

Fraction Catch

MATH CONCEPTS: proper and improper fractions, comparing fractions, equivalent fractions, number line, numerical order.

PLAYERS: any number.

EQUIPMENT: one set of double-six or double-nine dominoes.

How to Play

Remove the double-blank tile. Turn all remaining domino tiles face down on the table and mix them around to make the wood pile.

Turn up two tiles to start your number line. Place one domino as a proper fraction (numerator less than denominator) and the other as an improper fraction. A blank is a zero, and a fraction may never have zero in the denominator, so the blank always has to go on top. If the two tiles are equivalent—such as 1|1 and 5|5, or two tiles with blank sides—then mix one back into the wood pile and draw a replacement.

On your turn, draw one tile from the wood pile. Create a fraction (proper or improper) with the numbers on the two halves of the tile. Add your fraction to the number line as follows:

- ♦ If your fraction fits on either end of the number line, place it in the appropriate position.

- ♦ If your fraction fits between two already-placed fractions, you score a point. Capture the smaller of the two fractions for your personal score pile, and put your fraction in its place.

- ♦ If your fraction is equivalent to one already showing, set it aside in a discard pile. But first, check to see if you can flip your tile around to make a different fraction that might fit on the number line.

Play until all the dominoes are used or discarded. Whoever captures the largest score pile wins the game.

The number line has four fractions in place,
and then you draw the 2|4 tile.
Can you fit it in?

Variations

Allow players to draw a second tile if their first one won't play. Also, the discards may be mixed back into the wood pile, if you like. They might work the next time they are drawn, if someone has captured one of the equivalent tiles in the meantime.

STRATEGY VERSION: At the start of the game, players draw three tiles to form their hands. On your turn, play one of these tiles on the number line, and then draw to replenish your hand. If you have no play, discard one tile and draw a replacement.

QUICK CATCH: The first player to capture five tiles wins the game. Or play to some other agreed-upon target.

PREALGEBRA FRACTION CATCH: Allow fractions to be designated as negative numbers and placed below zero on the number line. You may

wish to place an acrylic gem or other token on the tile to remind everyone that it's negative.

FAREY FRACTIONS: (A cooperative or solitaire puzzle.) Turn all the tiles
up and remove the blanks and doubles. Make proper fractions. Keep
only the simplest form of each equivalent fraction: these lowest-term
fractions are the *Farey fractions.* Can you put them in sequence from
least to greatest? For an added challenge, try to space the fractions to
scale, as on a ruler. Do you notice a pattern?

History

I discovered this version of the Number Train game on John Golden's
Math Hombre blog. Golden created a special deck of fraction cards,
which you can download and print on paper or cardstock.[†]

The Farey fraction sequence is named for British geologist John
Farey, who did not discover it. A century later, the famous G. H. Hardy
disparaged Farey, writing that "he seems to have been at the best an
indifferent mathematician." Debra Borkovitz blogged about "Farey
Fraction Visual Patterns" and commented:

> *I rather like the idea that the Farey Sequences are named after
> someone who noticed a pattern and asked a question—and not even
> the first person to notice the pattern, ask the question, or provide the
> answer.*
>
> *As math teachers, we teach plenty of indifferent mathematicians
> who wake up when they experience the joy of discovering something
> that is new to them, not necessarily new to the whole world.[‡]*

---

† *mathhombre.blogspot.com/2012/01/fraction-catch.html*
‡ *debraborkovitz.com/2012/06/farey-fraction-visual-patterns*

My Nearest Neighbor

Math Concepts: comparing fractions, equivalent fractions, benchmark numbers, thinking ahead.

Players: two to four.

Equipment: math cards (two players need one deck, three or four players need a double deck).

How to Play

Deal five cards to each player. Set the rest of the deck face down in the middle of the table as a draw pile.

You will play four or more rounds:

◆ Closest to zero

◆ Closest to one

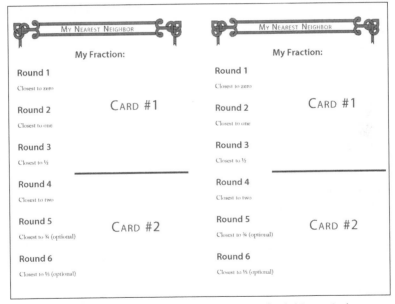

You don't need a game board, but you may find this reminder page from the *Multiplication & Fraction Printables* file useful. Cut the page in half and give one sheet to each player.[†]

† *tabletopacademy.net/free-printables*

- ◆ Closest to ½
- ◆ Closest to two
- ◆ Closest to ¾ (optional)
- ◆ Closest to ⅓ (optional)

In each round, players lay down two cards from their hand to make a fraction as close as possible (but not equal) to the target number. Draw two cards to replenish your hand.

The player whose fraction is nearest neighbor to the target collects all the cards played in that round. If there is a tie for closest fraction, the winners split the cards as evenly as they can, leaving any remaining card(s) on the table as a bonus for the winner of the next round.

After the last round, whoever has collected the most cards wins the game.

This fraction is close to one-half, but the player
could get closer. Can you see how?

History

A dream team of math teachers who call themselves the MathTwitter-Blogosphere (#MTBoS) created the Twitter Math Camp offline conference, which meets in a different city each year. At TMC 2014, participants played and tweaked math games, including Fraction War (page 49). Check out the TMC website for conference session archives and information about future meetings.[†]

TMC camper Christine Sullivan shared a favorite variation at A Sea of Math blog, which I adapted to make my version above.[‡]

Joshua Greene of Three J's Learning blog played the game with third- and fourth-grade students and suggested the current order for the target rounds, moving the hardest fractions to the end. He also liked a variation allowing fractions equal to the target number to win.

[†] *twittermathcamp.com*
[‡] *a-sea-of-math.blogspot.com/2014/08/my-closest-neighbor-estimating-with.html*

Decimal Product Game

MATH CONCEPTS: multiplying decimals, decimal place value.
PLAYERS: only two.
EQUIPMENT: printed game board, colored markers or a set of matching tokens for each player, two acrylic gemstones or other small tokens to mark the factors.

How to Play

The first player places a stone on any one of the factors at the bottom of the board. The second player places the other stone on a factor—the same or different—and then marks the product of those two numbers.

DECIMAL PRODUCT GAME					
.01	.02	.03	.04	.05	.06
.08	.09	.1	.12	.15	.16
.2	.25	.3	.4	.5	.6
.8	.9	1	1.2	1.5	1.6
2	2.5	3	4	5	6
8	9	10	12	15	16
20	25	30	40	50	100

1   2   3   4   5   10
.1   .2   .3   .4   .5

John Golden modified The Product Game's board to give students practice in multiplying decimal numbers.[†]

---

† *tabletopacademy.net/free-printables*

On each succeeding turn, a player moves just one stone to a new number and then marks the product of those two factors. If both players agree that all possible moves have already been colored in, the player whose turn it is may make a fresh start by moving both stones.

Whichever player marks four (or more) squares in a row—horizontal, vertical, or diagonal—wins the game. The squares must touch each other at edges or corners, with no gaps. If neither player can make four in a row, then the player who has the most sets of three in a row wins.

History

This variation on The Product Game (page 53) comes from John Golden's Math Hombre blog. He writes about his conversation with fifth-grade students:

> *We discussed how to get hundredths. I think they knew on some level it was tenths times tenths, but had a bit of the "we haven't been taught this yet" syndrome. I used the analogy of a dime being a tenth of a dollar, so what's a tenth of a dime?*
>
> *"A penny!"*
>
> *And what part of a dollar is a penny? How many does it take to make one dollar? It is terrific that they are used to seeing one cent written as .01.*[†]

† *mathhombre.blogspot.com/2011/03/product-game-again.html*

Fraction Product Game

MATH CONCEPTS: multiplying fractions, equivalent fractions.
PLAYERS: only two.
EQUIPMENT: printed game board, colored markers or a set of matching tokens for each player, two acrylic gemstones or other small tokens to mark the factors.

How to Play

The first player places a stone on any one of the factors at the bottom of the board. The second player places the other stone on a factor—the same or different—and then marks the product of those two numbers.

On each succeeding turn, a player moves just one stone to a new number and then marks the product of those two factors. If both players agree that all possible moves have already been colored in, the player whose turn it is may make a fresh start by moving both stones.

Whichever player marks four (or more) squares in a row—horizontal, vertical, or diagonal—wins the game. The squares must touch each other at edges or corners, with no gaps. If neither player can make four in a row, then the player who has the most sets of three in a row wins.

History

Have you figured out yet how much I love John Golden's blog? About this game, Golden writes:

> I launched the game by playing me vs. the class. Re-emphasizing that you only get to change one factor at a time, the goal is to get four in a row, and the new idea that there are equivalent fractions. If you multiply and get $\frac{6}{12}$, you can cover $\frac{1}{2}$, and vice versa.
>
> Then the students played pair vs. pair. At the end we summarized

1	½	⅔	⅓	¾	2/4
1/12	9/16	6/16	4/16	3/16	¼
2/12	3/24	2/24	1/24	2/16	5/6
3/12	5/24	1/36	25/36	1/16	2/6
4/12	10/24	5/36	18/36	10/18	1/6
5/12	15/24	1/18	2/18	5/18	3/8
6/12	1/9	2/9	4/9	1/8	2/8

½ ⅓ ⅔ ¼ 2/4
  ¾ 1/6 5/6 1

My printables file includes traditional and spiral forms for each of the Product Games. By putting the more difficult math facts in the valuable center squares, the spiral version encourages students to stretch their skills.

by discussing what they noticed about the game, and what they thought made for a good strategy.

I did point out to them that someone would tell them fraction division was hard, but they've already done it when they're figuring what to multiply ¾ by to get 6/12.[†]

_____

† *mathhombre.blogspot.com/2010/05/multiplying-fractions-times-three.html*

Playing to Learn Math

Diagnosis: Workbook Syndrome

WHETHER IT IS FICTION, BIOGRAPHY, news, sports, or even just the comics, most of us have something we read for enjoyment. But have you ever done math just for the fun of it, for the joy of discovery?

Many people would consider that a nonsense question. Mathematics has nothing to do with joy. Math is a chore to be endured, like cleaning the bathroom, one of those things that nobody likes but that has to be done.

Isn't it?

Well, no. At least, it doesn't have to be like that.

Our childhood struggles with school math gave most of us a warped view of mathematics. We learned to manipulate numbers and symbols according to what seemed like arbitrary rules. Most of us understood a bit here and a bit there, but we never saw how the framework fit together. We stumbled from one class to the next, packing more and more information into our strained memory, until the whole structure threatened to collapse. Finally we crashed in a blaze of confusion, some of us in high school algebra, others in college calculus.

Now that we are parents and teachers, we see the danger for our children. Many of us dread helping with our children's math homework, not knowing how to explain something that never made sense to us. Homeschoolers switch from one math program to another looking for a magic bullet. Classroom teachers follow the manual faithfully and hope for the best. Or they try a more creative route, scouring professional magazines, websites, and teacher blogs for activities and group projects to supplement the curriculum, hoping something will catch the students' imagination.

Some teaching philosophies recommend a strong focus on memorization in the early grades. Young children, they say, excel at memory work but cannot think logically, so conceptual explanations are wasted on them. Others argue that we must focus on understanding because rote memorization and speed drills can kill a child's interest in mathematics. Some rely on teaching rules and patterns, trusting that insight will follow as these become automatic. Still others push for plenty of hands-on experience that will allow students to draw their own conclusions about how numbers work.

Too often, discussions about math education, at least in America, devolve into a battle of stereotypes and straw-man arguments called the Math Wars.

In our confusion about how to teach mathematics, we have not yet found a way to protect our children from workbook syndrome. What, you may ask, is workbook syndrome? It is a distressing malady that afflicts children in public, private, and home schools across our country. A child suffering from this disease has learned to do calculations on a school math page but cannot make sense of numbers in word problems or in real life.

Mischievous Results of Poor Teaching

Educational pundits may try to blame one side or the other of the Math Wars, but the problem of workbook syndrome goes back at least to the nineteenth century. Victorian educator Charlotte Mason describes the symptoms:

> *There is no one subject in which good teaching effects more, as there is none in which slovenly teaching has more mischievous results.*

Multiplication does not produce the right answer, so the boy tries division; that again fails, but subtraction may get him out of the bog. There is no 'must be' to him—he does not see that one process, and one process only, can give the required result.

Now, a child who does not know what rule to apply to a simple problem within his grasp has been ill taught from the first, although he may produce slatefuls of quite right sums ... The child may learn the multiplication table and do a subtraction sum without any insight into the rationale of either. He may even become a good arithmetician, applying rules aptly, without seeing the reason of them.

I discovered a case of workbook syndrome in math club one afternoon, as I played Multiplication War with a pair of fourth-grade boys. They did fine with the small numbers and knew many of the answers by heart, but they consistently tried to count out the times-nine problems on their fingers. Most of the time, they lost track of what they were counting and ended up wildly wrong.

I stopped the game in midturn to teach a mental math technique: multiplying by nine is the same as multiplying by "ten minus one." Nine of anything is the same as ten of that thing, take away one of them. Nine books is ten books take away a book, and nine horses is ten horses take away one horse. With numbers, 9×6 is ten sixes take away one six, or $60 - 6 = 54$. Similarly, 9×8 is ten eights take away one eight, or $80 - 8 = 72$. It works for any number. For instance, 25×9 is ten twenty-fives take away one twenty-five, or $250 - 25 = 225$. By reducing the multiplication to a much simpler subtraction, this trick makes the times-nine table a cinch.

We spent a few minutes going through the times-nine facts together, just to practice the pattern.

$$1 \times 9 = 10 - 1$$
$$2 \times 9 = 20 - 2$$
$$3 \times 9 = 30 - 3$$
$$4 \times 9 = 40 - 4, \text{etc.}$$

Nine pencils is (10 − 1) pencils.

To my surprise, the older boy could not subtract without counting on his fingers. In several years of classroom training, he had not learned the number bonds, the pairs of numbers that make ten. No, that can't be true. I am sure he had learned them, probably in kindergarten, but his teachers had never led him to see how these simple facts could help him solve problems, so he had just forgotten them. When I probed further, I found he could not mentally add ten to a two-digit number.

When I gave him a pencil and wrote the numbers on paper, the boy knew how to follow the procedures for adding and subtracting into the thousands and beyond. He'd been taught the standard arithmetic *algorithms*—the traditional set of abstract, multistep rules—yet he had almost no understanding of how numbers work. He had been shafted by several years of poor instruction dished out by teachers who themselves did not understand math.

The Struggle for Balance

We all know that number skills are important to our children's future. As students work their way through elementary and middle school arithmetic, they must:

♦ Understand number concepts, the basic principles of how numbers work together, such as addition, division, or the distributive property.

- ◆ Memorize the *math facts*, the simple relationships between small numbers, such as 3 + 5 = 8 and 7 × 2 = 14.

- ◆ Learn to apply these concepts and facts in an ever-growing variety of situations.

In a balanced math education, both components—conceptual understanding and knowledge of basic facts—will grow together. Having number facts in memory makes calculation easier, which allows the students to concentrate on whatever puzzle or problem they are trying to solve. This enables them to add new layers to their understanding and begin to appreciate the more interesting concepts of mathematics. And this new understanding in turn brings new light to the math facts, making them easier to recall.

Unfortunately, I've never found a math program that mixes these components perfectly for my children. My solution has been to pick our math program based on how well it helps my children learn the foundational concepts, knowing I can build in lots of number practice by playing games.

Whatever math curriculum your children use, do not be satisfied with mere pencil-and-paper competence. To prevent workbook syndrome, help your children develop mental math skills. Mental calculation forces a child to understand numbers, because the techniques that let us work with numbers in our heads reinforce the fundamental concepts of arithmetic.

For instance, when my math club boys forgot the times-nine facts, I taught them a technique based on the *distributive property*, one of the most basic principles in math. You can think of it as the shopping bag rule: if you buy fruit in mixed bags, you can take them home and separate the pieces of fruit according to their types. Imagine buying six bags, and each bag contains three apples and two pears. You could say you bought six bags with five pieces of fruit in each bag, 6 × 5 = 30 pieces of fruit. But looking at it by types, you could also say you bought 6 × 3 = 18 apples and 6 × 2 = 12 pears. When you put that information into a single equation, the parentheses act like grocery bags:

Don't think of math rules as mere abstractions. They reflect common sense about how the real world works.

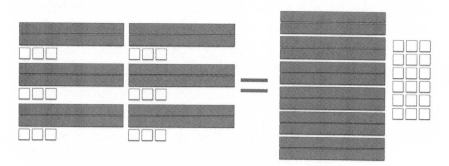

Numbers act just like apples and pears. Six sets of two tens and three ones (6 × 23) is the same as 6 × 2 tens and 6 × 3 ones.

Fruit in mixed bags = fruit sorted by type
$$6 \times (3 + 2) = (6 \times 3) + (6 \times 2)$$

Or if we let a, b, and c represent any numbers:
$$a \times (b + c) = (a \times b) + (a \times c)$$

Be wary of teaching finger tricks, catchy rhymes, or other mnemonic aids that hide what the numbers are doing. Such things burden your child's memory without increasing understanding. For more information about math tricks to avoid, download Tina Cardone's free ebook *Nix the Tricks: A Guide to Avoiding Shortcuts That Cut Out Math Concept Development.*[†]

You can build mental calculation skills by doing math homework orally. To work math problems in their heads, children have to learn how to take numbers apart and put them back together. They figure out ways to simplify calculations and learn to recognize common patterns. The numbers become, in a sense, their friends.

Oral work has another advantage: young children need not be limited by their still-developing fine motor skills. My sons, especially, could advance quickly through math topics that they would never have had the patience to write out. As students progress to more difficult problems, they may wish to have scratch paper or a lap-size whiteboard and colorful markers handy. Even then, however, we do as much work as we can mentally.

As veteran teacher Ruth Beechick writes, "If you stay with meaningful mental arithmetic longer, you will find that your child, if she is average, can do problems much more advanced than the level listed for her grade. You will find that she likes arithmetic more. And when she does get to abstractions, she will understand them better."

---

† *nixthetricks.com*

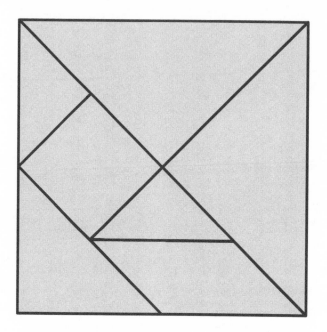

Have your children ever played with tangrams?
Try applying Professor Triangleman's Standards for
Mathematical Practice to this mental math puzzle:
What fraction is each tangram piece, compared to the whole square?

Just like the games we play, the fun in learning mathematics is in the challenge.

—ERLINA RONDA

Conclusion: Master the Math Facts

WHEN IT COMES TO PRACTICING number facts, many children think math is spelled "b-o-r-i-n-g." Worksheets are tedious, flash cards make them groan, and even the latest computer game is a yawner. School supply websites feature a variety of educational products designed to make the process easier, from rods and blocks to Math Fact Bingo. Over the years, I have spent hundreds of dollars on products designed to help my children learn math. Maybe a multiplication coloring book with silly stories will help this year, or perhaps we should try a CD of skip-counting songs?

Even children who understand arithmetic well may struggle to master the basic number facts. Learning to understand math is a conceptual task, but learning the math facts is more like rote memory work. Yet rote memory is not enough. A student may recite the times tables perfectly and still be reduced to counting on fingers in the middle of a long-division problem.

Training one's mind to recall answers when needed is a lot like learning to type. It comes in stages.

Stage One: Hunt and Peck

In typing, we understand that we have to push down the proper key to get the letter we want, but it may take us a few minutes to find that key. In math, this is the manipulative or counting-on-fingers stage.

Stage Two: Slow but Steady

Now we have learned that each finger controls certain keys, but we have to think about whether "c" is up or down from the home row. In math, students understand the concepts behind each math fact, but they still count by five to calculate 5×7.

Stage Three: Automatic Response

Professional typists look at a word on the paper they are copying, and their fingers automatically hit the proper sequence of keys. Typing has become a reflex. A math student who has reached this stage can see 2×5 on a worksheet and instantly think "10."

Of course, we do not progress evenly from one stage to the next. As a typist, I work primarily in stage two, but simple words (*the* or *and*) are automatic, while I still hunt and peck the numbers and unusual forms of punctuation. For our students, progress in learning math will come the same, slow way. They may know instantly that 3×5 is the same as 15, while they still count on their fingers to solve monsters like 8×6.

Also, notice that not all typists reach the automatic stage. I have a friend who can type more than one hundred words per minute, almost as fast as she can think. I can type around thirty words per minute, which is about as fast as I can think, too. Would I like to type faster? Sure, but not enough to work at it. I will never be a medical transcriptionist, but I can type well enough for email.

In the same way, not every student will reach the automatic stage with all the number facts. Most of us still struggle with remembering a few of them as adults, often the times-seven or times-eight facts. As long as we know how to figure out the ones we cannot recall, we will survive.

As with typing, there is only one way to reach the automatic stage: practice, practice, practice. The student must calculate the number relationships over and over and over, so many times that the correct response becomes a reflex. Practice makes permanent.

Thankfully, with the games in the *Math You Can Play* series, practicing the math facts can be fun.

A Strategy for Learning

There is no perfect teacher. There is only you and me, and we have no superpowers. We can't save the world or solve the latest crisis of educational policy.

But we can help our children learn to do mental math. We can encourage them to practice strategic thinking and develop problem-solving skills. We can prevent (or treat) math anxiety and build a positive attitude toward learning.

So what are we waiting for? Let's play some math!

Resources and References

Game-Playing Basics, From Set-Up to Endgame

WHEN I WAS A CHILD, I assumed that whatever I knew was common knowledge and anything I believed was common sense. Now I have grown up enough to realize how very much I do not know. So I understand how confusing new ideas (or new games) can be.

Below I summarize everything I can think of that might be assumed-but-never-explained about playing games in the *Math You Can Play* series. If you have a question I didn't answer, please send me an email.[†]

Math Concepts

Most of the *Math You Can Play* games build your students' skill at working with numbers in their heads. Some games focus on one or two concepts, while others cover a wide range of ideas. The latter are not necessarily better than the former.

Players

Almost all of the games are designed for two or more competing players, but a few also work as solitaire games. If a game relies on chance, it may not matter how many people are playing. But the more strategy involved in a game, the more likely it will work best as a two-player battle of minds.

When playing with a larger group, it may work better to split up and play separate games so players don't have too much idle time between their turns. To wait patiently can be difficult for an adult, so we shouldn't be surprised that it's hard for children.

If you are playing with a wide range of skill levels, avoid games that rely on speed or change them to allow each player adequate time to think. In some

[†] *LetsPlayMath@gmail.com*

cases, you may allow extra turns to the younger students. For instance, in Concentration (Memory), you might let younger children turn up three cards instead of the usual two, giving them a better chance to find a pair that match.

Who Goes First?

Sometimes there is an advantage to going first in a game, so I often let the youngest player go first. You can randomize the turns by letting each player throw a die or draw one card from the deck, and then whoever gets the highest number goes first. In multiplayer card games, whoever gets the lowest number is the dealer, and the player sitting to the dealer's left goes first.

Shuffle and Cut

In card or domino games, the players must mix the cards or tiles to ensure randomness. Any player may shuffle, but in card games, the dealer has the right to shuffle last. Players should not try to sneak a peek during the shuffle, and to take advantage of an accidentally revealed card is cheating.

The riffle shuffle (in which a deck of cards is split in half and then the halves are interlaced) takes plenty of practice, but it is the fastest way to randomize the cards. It takes about seven riffles to fully shuffle a deck. If you've never seen a riffle shuffle, search YouTube for a video that shows the technique.[†]

For young children, the easiest way to shuffle is domino style. Spread all the cards face down on the table, mix them around, and then stack them up again without looking.

If only the dealer shuffles the cards, it is polite to offer another player (usually whoever is to the dealer's right, opposite the direction of the deal) the chance to cut the deck. The player splits the deck into two parts, with at least four cards in each part, places the top part on the table without looking at it, and then stacks the other part on top of it. Thus neither player nor dealer can know the exact position of any card.

Deal and Rotation of Play

In many card games, one person—the dealer—will hand out cards to each player in turn, going around the table in the direction of play. The dealer may

† *youtu.be/3oabnbtJRNQ*

The riffle shuffle is an efficient way to randomize a deck of cards.

give one card at a time, or two or more at once, but should deal the same to every player. Out of politeness and to avoid putting other players at a disadvantage, everyone should wait until all cards are dealt before picking up and looking at their hands.

My family plays by the tradition common in the United States that the deal and the players' turns go to the left (clockwise) around the table. I have read that many countries do the opposite, rotating play to the right. For the games in this book, direction does not matter, so use whichever seems comfortable to you.

Hand vs. Round vs. Game

The cards a player holds are called his or her hand. These are normally kept hidden from the other players until used in the game. Children often need to be reminded to hold their hands close to their bodies so that the other players do not see their cards.

One complete section of a game, where every player has a turn or chance to play, may be called a hand or a round. Sometimes the terms are interchangeable, but for more complicated games it may take several hands to make a round and several rounds to finish a complete game.

Draw Pile (Stock) and Discard Pile

In games where players will need to refresh their hands, the rest of the deck (after cards have been dealt) is turned face down and placed on the table where everyone can reach. This is the *stock* or *draw pile*.

The unwanted cards from the players' hands are often turned face up next to the draw pile, either as a single stack or fanned out so all are visible. In some games, these discards may be available for other players to use in subsequent turns.

In many games for young children, a *fishing pond* is used in place of a draw pile. Turn all the cards face down and spread them out to form a roughly circular area where all players can reach. On their turns, players may choose any card. Discards should be mixed back into the pond before the next player's turn, so that nobody can remember their location.

Misdeal and Other Irregularities

When playing with children, you can almost guarantee that misdeals or exposed cards will happen. Some traditional games specify harsh penalties for such irregularities—think of old Western movies, where a game of poker could break into a bar fight or shootout over a simple mistake. In a family game, we can be more lenient. Our children must learn that cheating ruins the game for everyone, but there is no shame in an unintended error.

The dealer may give the wrong number of cards to a player or expose a card that other players are not supposed to see. Just fix the misdeal in a way that seems fair all around—for instance by mixing the offending cards back into the deck or by reshuffling and starting over. If the players have looked at their hands before realizing they have too many cards, no one should choose which of his or her own cards to give back. Players can fan out their hands and let the dealer or another player who can't see the cards pick and discard the extras.

In the same way, anything wrong that happens in a game should be resolved in such a way as seems fair to every player. For example, if someone plays a card out of turn or makes an illegal play, the exposed card should stay face up on the table and be used at the next legal opportunity. Or if the deck is bad (perhaps the players discover that a few cards are missing), the current hand should start over with a new deck, but any points from previous rounds should stand.

Keeping Score

At our house, we often play for the next deal, rather than for points. My children enjoy having control over the game, so getting to deal is a treat for them. When we do play for points, the kids love to use poker chips to keep score: white = 1 point, red = 5 points, and blue = 10 points.

You could let your children practice money skills by using coins to keep score. Give one penny per point, with players trading in for higher coins as they progress, and the first player to collect $1 (or $5 or $10) wins the game.

Or you may use face cards and jokers as tallies in games where the winner of each hand gets a point. Give one tally card per point until they are gone, and whoever collects the most cards is the champion. Since there are three face cards in each of the four suits, this will make twelve hands, or fourteen with both jokers.

Or Try a Cribbage Board

You can often pick up cards, dice, dominoes, and poker chips at garage sales for next to nothing. A rarer discovery worth grabbing if you see one is a cribbage board (sometimes called a *crib board*). You don't have to know how to play cribbage to find this useful. It can be a great way to record points in many different games.

Keep score using two small pegs in leapfrog fashion. Each hole represents a point, and the holes are arranged in groups of five for easy counting. To record your first score, count that many holes and place your first peg. For the second score, leave the first peg where it is and count beyond it, placing

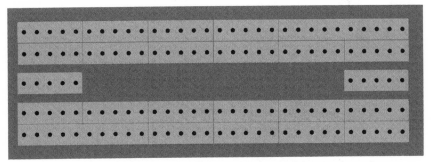

A 60-point cribbage board for two players.
Move the pegs away from yourself up your first row of thirty holes
and then back down the second row. Use the holes in the middle
for longer games, to count your trips around the board.

the second peg at your new total. For each succeeding set of points, leave the farthest-advanced peg in place to guard against losing count, and jump the other peg past it to the new total. The first player to peg out—that is, to reach the last hole—wins the game.

If you buy it used, your cribbage board may have lost its pegs. You can snip the sharp ends off round toothpicks or use wooden matchsticks as makeshift cribbage pegs.

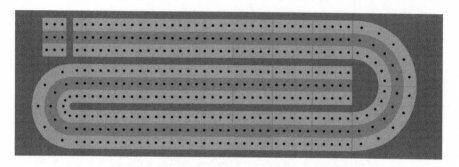

A three-player cribbage board for a game of 120 points.
Follow your own line of holes around the loop to the
end. If you use the three extra holes to count trips
around the board, you can play longer games.

A Few of My Favorite Resources

IF YOU KNOW OF A fantastic math games resource I missed, please send me an email. I appreciate your help!

Best-Loved Books

Most of these books should be available through your local library or via inter-library loan. Check for recreational games in the 793–795 range in the Dewey decimal system, and look for elementary education games at 372.

Camp Logic by Mark Saul and Sian Zelbo

Family Math by Jean Kerr Stenmark, Virginia Thompson, and Ruth Cossey

Games for Math by Peggy Kaye

Games with Pencil and Paper by Eric Solomon

Hexaflexagons and Other Mathematical Diversions and other books by Martin Gardner

Math for Smarty Pants and *The 'I Hate Mathematics!' Book* by Marilyn Burns

Math Games and Activities from Around the World and other books by Claudia Zaslavsky

Mathematical Activities: A Resource Book for Teachers and *The Amazing Mathematical Amusement Arcade* and other books by Brian Bolt

Moebius Noodles: Adventurous Math for the Playground Crowd by Yelena McManaman and Maria Droujkova

Playing with Math: Stories from Math Circles, Homeschoolers, and Passionate Teachers edited by Sue VanHattum

Online Games and Resources

The Internet overflows with a wide-ranging assortment of math websites. The list at my blog is much longer than this, and the "good intentions" folder of links I hope to add someday is longer still.

AMBLEWEB FUNCTION MACHINE: Choose the type of problem you want to guess, or go random for more challenge. My math club kids love function machines.
amblesideprimary.com/ambleweb/mentalmaths/functionmachines.html

CUT THE KNOT INTERACTIVE: "Mathematics Miscellany and Puzzles," one of my all-time favorite sites. See also "Math Games and Puzzles, A Short Illustrated List."
cut-the-knot.org
cut-the-knot.org/games.shtml

DAILY TREASURE: Solve the logic puzzle to find the hidden gold.
4chests.blogspot.com

GAMES AND MATH AND PAM SOROOSHIAN ON DICE: Math itself is a game we play.
sandradodd.com/math/pamgames
sandradodd.com/math/pamdice

HEAD HUNTERS GAME: A bloody fun game for the Viking in all of us. If you enjoy that one, try the other math tricks and games at Murderous Maths.
murderousmaths.co.uk/games/headhunt/headhunt.htm
murderousmaths.co.uk

INCOMPETECH: Free online graph paper PDFs galore for any math game.
incompetech.com/graphpaper

KENKEN FOR TEACHERS: A playful way to practice arithmetic.
kenkenpuzzle.com/teachers/classroom

Many interactive math websites require Java or Adobe Flash. Unfortunately, both programs can also be used by hackers to break into your computer or do other nasty stuff. Make sure you have the most recent versions of each program, and keep your security settings up to date.
java.com
java.com/en/download/help/java_blocked.xml
java.com/en/download/faq/exception_sitelist.xml
adobe.com/products/flashplayer.html

MATH PICKLE: Videos introduce fun and challenging printable games and puzzles for K–12 students. Can your students solve the $1,000,000 problems?
mathpickle.com

MATH PLAYGROUND: My favorite site for a variety of math games.
mathplayground.com

MATH WORKSHEET SITE: My personal favorite hundred chart generator.
themathworksheetsite.com

MATHEMATICAL GAMES AND RECREATIONS: "The whole history of mathematics is interwoven with mathematical games which have led to the study of many areas of mathematics."
www-groups.dcs.st-and.ac.uk/~history/HistTopics/Mathematical_games.html

NATURAL MATH: Plenty of ideas for sharing rich math experiences with your children.
naturalmath.com

NRICH.MATHS.ORG: A wonderful source of math games and activities for all ages, with a theme that changes each month.
nrich.maths.org/public/index.php

PAGAT.COM: A collection of card game rules and variations from around the world.
pagat.com

RECREATIONAL MATHEMATICS: Games, art, humor, and more.
mathworld.wolfram.com/topics/RecreationalMathematics.html

RUSH HOUR ONLINE: A fun logic puzzle game.
puzzles.com/products/RushHour/RHfromMarkRiedel/Jam.html

SCRATCH: A programming language that makes it easy for students to create interactive stories, animations, games, music, and art.
scratch.mit.edu

SET DAILY PUZZLE: Logic puzzle game for all ages.
setgame.com/set/puzzle_frame.htm

I checked all these website links in May 2016, but the Internet is volatile. If a website disappears, you can run a browser search for the author's name or article title. Or enter the web address at the Internet Archive Wayback Machine.
archive.org/web/web.php

Taxicab Treasure Hunt: A game based on the non-Euclidean geometry of city streets.
learner.org/teacherslab/math/geometry/shape/taxicab/index.html

Ultimate List of Printable Math Manipulatives & Games: A treasure list from one of my favorite homeschooling blogs.
jimmiescollage.com/2011/04/ultimate-list-of-printable-math-manipulatives-games

Board Games for Family Play

Board games are a celebration of problem solving, and problem solving is at the heart of a quality mathematics education. The mathematics might be hidden, but I guarantee you that it will be there.

—Gordon Hamilton

In addition to the classics of strategy—backgammon, chess, mancala, Othello/Reversi, Pente, and so on—your family can enjoy (and learn from) many modern games:

Blokus: Strategy game for up to four players.

Carcassonne: Lay down your tiles to create a landscape based on southern France.

Citadels: Bluffing, deduction, and city-building set in a medieval world.

Dvonn: An abstract strategy game based on moving pieces to make the largest stack.

For Sale: Buy and sell real estate to amass your fortune.

Forbidden Island: Capture four sacred treasures from the ruins of this perilous paradise.

King of Tokyo: Mutant monsters, gigantic robots, and other aliens vie for the right to rule the city.

Labyrinth: Find a path to collect your treasures, but watch out—the maze shifts and changes on every turn.

Lost Cities: Explore the world in search of ancient civilizations.

Love Letter: A short, simple game that combines luck and strategy.

MEMOIR '44: Test your strategic skills as you refight the battles of World War II.

MR. JACK: Jack the Ripper is loose in Whitechapel, and it's up to you to stop him.

MUNCHKIN QUEST: Explore the dungeon and battle monsters for power and treasure.

POWER GRID: Acquire raw materials, upgrade your power plants, and expand your network to more cities.

QUARTO: Four-in-a-row strategy game.

QUIRKLE: Strategically match colors and shapes to build up your score.

QUORIDOR: Move your pawn through the maze, and block the other players.

SET: Visual perception card game.

SETTLERS OF CATAN: A trading and building game set in a mythical world.

7 WONDERS: Lead an ancient civilization as it rises from its barbaric roots to become a world power.

SMASH UP: Easy to learn, fun to play, and always different.

SPLENDOR: As a Renaissance merchant, you must acquire mines and transportation, hire artisans, and woo the nobility.

STONE AGE: Gather resources to feed and shelter your tribe.

TICKET TO RIDE: A cross-country train adventure. How many cities can you visit?

ZEUS ON THE LOOSE: Use addition, subtraction, and strategic thinking to capture the runaway god.

ZOOLORETTO: Plan carefully to attract as many visitors as possible to your zoo.

Quotes and Reference Links

I LOVE QUOTATIONS. EVERYTHING I could ever want to say has probably been said sometime by someone else (who did not think of it first, either). At least a few of those people had a wonderful way with words.

Some of the quotations in this book are from my own reading. Others are gleaned from two websites I visit often to browse: Furman University's Mathematical Quotation Server and the Mathematical and Educational Quotation Server at Westfield State College.[†]

All the website links below were checked in May 2016 (dates of original publication are included if the website provided them), but the Internet is volatile. If a website disappears, you can run a browser search for the author's name or article title. Or try entering the web address at the Internet Archive Wayback Machine.[‡]

ANONYMOUS. "We do not stop playing because we grow old..." Attributed to Benjamin Franklin, Oliver Wendell Holmes, and George Bernard Shaw, among others. Choose your favorite sage.

BEECHICK, RUTH. "If you stay with meaningful mental arithmetic..." from *An Easy Start in Arithmetic (Grades K–3)*, Arrow Press, 1986.

BENNETT, ALBERT BRADLEY, JR. Fraction Bars website. *fractionbars.com*

BERLEKAMP, ELWYN R, JOHN H CONWAY, AND RICHARD K GUY. *Winning Ways for Your Mathematical Plays*, A. K. Peters Ltd., 4 vols., 2001–2004.

BLUM-SMITH, BEN. "The most effective and powerful way..." from a comment on "What Must Be Memorized?" Math Mama Writes blog, May 26,

---

† *math.furman.edu/~mwoodard/mquot.html*
westfield.ma.edu/math/faculty/fleron/quotes
‡ *archive.org/web/web.php*

2010. Blum-Smith writes about teaching and learning mathematics at his Research in Practice blog.
mathmamawrites.blogspot.com/2010/05/what-must-be-memorized.html
researchinpractice.wordpress.com

BOALER, JO. "How Close to 100?" YouCubed website. Boaler, a Stanford University professor of mathematics education, produces a variety of resources to encourage students and parents who struggle with math.
youcubed.org/task/how-to-close-100

BOGOMOLNY, ALEXANDER. "Math Games and Puzzles: A Short Illustrated List," Cut the Knot website.
cut-the-knot.org/games.shtml

BORKOVITZ, DEBRA K. "Farey Fraction Visual Patterns," Out of the Math Box blog, June 25th, 2012. Borkovitz serves as an associate professor of mathematics at Wheelock College, helping train the next generation of elementary school teachers.
debraborkovitz.com/2012/06/farey-fraction-visual-patterns

BREWER, CHRIS. "Leap Frog Cover-up," shared by Alice P. Wakefield in *Early Childhood Number Games: Teachers Reinvent Math Instruction,* Allyn & Bacon, 1998.

BROADBENT, F W. "Contig: A Game to Practice and Sharpen Skills and Facts in the Four Fundamental Operations," *The Arithmetic Teacher,* May 1972.
jstor.org/stable/41188047
maconpiattroe.org/vimages/shared/vnews/stories/54c95745dbc67/Contig%20Rules.pdf

BURNS, MARILYN. *About Teaching Mathematics,* 3rd ed., Math Solutions Publications, 2007.

—. "The Game of Pathways," Marilyn Burns Math Blog, March 10, 2016.
marilynburnsmathblog.com/wordpress/the-game-of-pathways

CARDONE, TINA. *Nix the Tricks: A Guide to Avoiding Shortcuts that Cut Out Math Concept Development,* self-published, 2013.
nixthetricks.com

CHIALVO, FEDERICO. "Multiplication Tic Tac Toe," Art of Math Studio blog, August 30, 2013. Chialvo is a math specialist in the San Francisco Bay area at a K–8 independent school.
artofmathstudio.wordpress.com/2013/08/30/multiplication-tic-tac-toe

—. "Revisiting Multiplication Tic-Tac-Toe: Common Factors and Multiples," Art of Math Studio blog, October 31, 2013.
artofmathstudio.wordpress.com/2013/10/31/revisiting-multiplication-tic-tac-toe-common-factors-and-multiples

CHILDREN'S TELEVISION WORKSHOP. "But Who's Multiplying?" *Square One TV,* first season, 1987. Episodes available on YouTube.
tv.com/shows/square-one-tv/episodes
youtube.com/watch?v=p7C2w2WAeVU
youtube.com/watch?v=V-sBHbQBTgM
youtube.com/watch?v=iB9ff_NYJoY

DANIELSON, CHRISTOPHER. *Talking Math with Your Kids,* self-published, 2013. With his blogs and other projects, Danielson helps parents and teachers understand how children understand math.
talkingmathwithkids.com
christopherdanielson.wordpress.com

—. "One is one … or is it?" TedEd Lessons Worth Sharing website, 2012.
ed.ted.com/lessons/one-is-one-or-is-it

—. "Prof. Triangleman's Abbreviated List of Standards for Mathematical Practice," from "Ginger Ale (Also Abbreviated List of Standards for Mathematical Practice)," Overthinking My Teaching blog, December 10, 2013.
christopherdanielson.wordpress.com/2013/12/10/ginger-ale-also-abbreviated-list-of-standards-for-mathematical-practice

DEVLIN, KEITH. "Mastery in this context means…" from "Wanted: A Mathematical iPod," Devlin's Angle blog, June 2011. Devlin is a mathematician at Stanford University and the author of many books and several blogs.
maa.org/external_archive/devlin/devlin_06_11.html

—. "It Ain't No Repeated Addition," Devlin's Angle, June 2008.
maa.org/external_archive/devlin/devlin_06_08.html

—. "It's Still Not Repeated Addition," Devlin's Angle, July-August 2008.
maa.org/external_archive/devlin/devlin_0708_08.html

—. "Multiplication and Those Pesky British Spellings," Devlin's Angle, September 2008.
maa.org/external_archive/devlin/devlin_09_08.html

—. "What Exactly is Multiplication?" Devlin's Angle, January 2011.
maa.org/external_archive/devlin/devlin_01_11.html

DONNE, JOHN. "No man is an island…" from "Meditation 17," *Devotions upon Emergent Occasions,* 1623; excerpt available at Wikisource, full text at Project Gutenberg.
en.wikisource.org/wiki/Meditation_XVII
gutenberg.org/ebooks/23772

ERNEST, JAMES. "Gold Digger," Cheapass Games website. Printable rules and playing cards.
cheapass.com/free-games/gold-digger

FINKEL, DAN. "A Game to End All Times Tables Drills: Damult Dice," Math 4 Love blog, October 1, 2010. Finkel runs workshops on mathematics education and is a regular contributor to *The New York Times* Numberplay blog.
mathforlove.com/2010/10/a-game-to-end-all-times-tables-drills-damult-dice

GASKINS, DENISE. *Let's Play Math: How Families Can Learn Math Together, and Enjoy It,* Tabletop Academy Press, 2016.

—. "Fractions: ⅕ = ⅒ = ⅛₀ = 1?" Let's Play Math blog, August 13, 2014.
denisegaskins.com/2014/08/13/fractions-15-110-180-1

—. "How to Conquer the Times Tables," Let's Play Math blog, 2011.
denisegaskins.com/tag/times-table-series

—. "If It Ain't Repeated Addition, What Is It?" Let's Play Math, July 1, 2008.
denisegaskins.com/2008/07/01/if-it-aint-repeated-addition

—. "Number Game Printables Pack" and "Multiplication & Fraction Printables," Tabletop Academy Press website.
tabletopacademy.net/free-printables

—. "PUFM 1.5 Multiplication," Let's Play Math blog, 2012.
denisegaskins.com/2012/07/16/pufm-1-5-multiplication-part-1
denisegaskins.com/2012/10/02/pufm-1-5-multiplication-part-2

—. "Understanding Math: Multiplying Fractions," Let's Play Math blog, December 17, 2015.
denisegaskins.com/2015/12/17/understanding-math-fraction-multiplication

—. "What's Wrong with 'Repeated Addition'?" Let's Play Math, July 28, 2008.
denisegaskins.com/2008/07/28/whats-wrong-with-repeated-addition

GOLDEN, JOHN. "Be careful! There are a lot of useless games…" from "Math Games for Skills and Concepts" handout, Math Hombre blog. Golden helps train future math teachers as an associate professor at Grand Valley State

University, and trains the rest of us through the posts on his blog.
faculty.gvsu.edu/goldenj/GameshandoutHS.pdf

—. "Area Battle," Math Hombre blog, October 7, 2011.
mathhombre.blogspot.com/2011/10/area-battle.html

—. "Decimal Point Pickle," Math Hombre blog, May 30, 2010.
mathhombre.blogspot.com/2010/05/decimal-point-pickle.html

—. "Fraction Catch," Math Hombre blog, January 28, 2012.
mathhombre.blogspot.com/2012/01/fraction-catch.html

—. "Games Reference Page," Math Hombre blog.
mathhombre.blogspot.com/p/games.html

—. "Multiplying Fractions, Times Three," Math Hombre blog, May 13, 2010.
mathhombre.blogspot.com/2010/05/multiplying-fractions-times-three.html

—. "Multiplying Game Possibilities," Math Hombre blog, April 22, 2012.
mathhombre.blogspot.com/2012/04/multiplying-game-possbilities.html

—. "Product Game … Again!" Math Hombre blog, March 25, 2011.
mathhombre.blogspot.com/2011/03/product-game-again.html

GREENE, JOSHUA. "Dots and Boxes Variation," Three J's Learning blog, February 16, 2016. Greene blogs about exploring and playing with math as a family and in the classroom.
3jlearneng.blogspot.com/2016/02/dots-and-boxes-variation.html

—. "My closest neighbor," Three J's Learning blog, 2017.
3jlearneng.blogspot.com/2017/01/my-closest-neighbor-fraction-game.html
3jlearneng.blogspot.com/2017/01/closest-neighbor-one-on-one.html
3jlearneng.blogspot.com/2017/02/perfect-play-for-my-closest-neighbor.html

—. "Times Square Variations," Three J's Learning blog, November 18, 2015.
3jlearneng.blogspot.com/2015/11/times-square-variations-math-games.html

HAMILTON, GORDON. "Board games are a celebration…" from "Commercial Games," YouTube video, December 25, 2011. Hamilton posts games and activity ideas for all ages at his Math Pickle website.
youtu.be/J8geFOkOUbU
mathpickle.com

HARDY, G H AND E M WRIGHT. "He seems to have been at the best…," from *An Introduction to the Theory of Numbers*, Clarendon Press, 1938.

HAYAKAWA, SAMUEL ICHIYE. "The symbol is not the thing symbolized…" from *Language in Thought and Action*, 5th ed., Harcourt, 1990.

Huff, Darrell. *How to Lie with Statistics*, Penguin Books, 1973. Available at the Internet Archive.
archive.org/details/HowToLieWithStatistics

Kamii, Constance, and Catherine Anderson. "Multiplication Games: How We Made and Used Them," *Teaching Children Mathematics*, November 2003. If you don't have access to the TCM archive, check Nicora Placa's article "3 Games to Help Students Master Multiplication Facts."
nctm.org/Publications/teaching-children-mathematics/2003/Vol10/Issue3/
    Multiplication-Games_-How-We-Made-and-Used-Them
nicoraplaca.com/3-games-improve-multiplication-fact-fluency

—with Leslie Baker Housman. *Young Children Reinvent Arithmetic: Implications of Piaget's Theory*, 2nd ed., Teachers College Press, 2000.

—with Linda Leslie Joseph. *Young Children Continue to Reinvent Arithmetic, 2nd Grade: Implications of Piaget's Theory*, 2nd ed., Teachers College Press, 2004.

—with Sally Jones Livingston. *Young Children Continue to Reinvent Arithmetic, 3rd Grade: Implications of Piaget's Theory*, Teachers College Press, 1994.

Kawas, Terry. "Contig Jr.," MathWire website, 2009.
mathwire.com/games/contigjr.pdf

Kaye, Peggy. "Children learn more math and enjoy math more..." from *Games for Math*, Pantheon Books, 1988. If you're homeschooling, be sure to check out the other books in Kaye's *Games for...* series.

King, Colleen. "Number Bond Fractions," Math Playground website.
mathplayground.com/number_bonds_fractions.html

Leo, Lucinda. "With any curriculum there is the temptation..." from "Things I've Learned About Homeschooling," Navigating by Joy blog, December 10, 2013. Leo is an English mom who blogs about her family's unschooling adventures.
navigatingbyjoy.com/2013/12/10/3-things-ive-learned-homeschooling-2013

Mason, Charlotte. "There is no one subject..." from *Home Education*, 5th ed., 1906 (originally published 1886); available at Internet Archive. Mason encouraged parents to focus on word problems that build reasoning skills, to emphasize mental work over written sums, and to allow free access to manip-

ulatives as long as the child found them helpful.
archive.org/details/homeeducationser01masouoft

MathTwitterBlogosphere (#MTBOS). An informal group of online math teachers who organize the offline Twitter Math Camp and produce other creative math resources.
mathtwitterblogosphere.weebly.com/cool-things-weve-done-together.html
twittermathcamp.com

McLeod, John. "Card Game Rules: Card Games and Tile Games from Around the World," Pagat website. Pagat is a collection of card game rules and history tidbits.
pagat.com

—. "Rummy (Basic)," Pagat website.
pagat.com/rummy/rummy.html

—. "Twenty-Four," Pagat website.
pagat.com/adders/24.html

Meyer, Dan. "Tiny Math Games," dy/dan blog, April 16, 2013.
blog.mrmeyer.com/2013/tiny-math-games

Nctm Illuminations Team. "Product Game," Illuminations website.
illuminations.nctm.org/Activity.aspx?id=4213

Nunes, Terezinha, and Peter Bryant. *Children Doing Mathematics,* Wiley-Blackwell, 1996.

Nrich Team. "Stop or Dare," Nrich Enriching Mathematics website.
nrich.maths.org/1193

Orlin, Ben. "Tic-Tac-Toe Puzzles (and the Difference Between a Puzzle and a Game)," Math with Bad Drawings blog, November 18, 2013. Math teacher Orlin's blog is delightful and full of insight—well worth following.
mathwithbaddrawings.com/2013/11/18/
   tic-tac-toe-puzzles-and-the-difference-between-a-puzzle-and-a-game

—. "Ultimate Tic-Tac-Toe," Math with Bad Drawings blog, June 16, 2013.
mathwithbaddrawings.com/2013/06/16/ultimate-tic-tac-toe

Pierce, Rod, and Others. "Mathematical Models," Maths Is Fun website.
mathsisfun.com/algebra/mathematical-models.html

Placa, Nicora. "3 Games to Help Students Master Multiplication Facts,"

Bridging the Gap blog, December 16, 2013.
nicoraplaca.com/3-games-improve-multiplication-fact-fluency

PLATO. "There should be no element of slavery in learning..." from *The Republic.* Quoted at the Mathematical and Educational Quotation Server at Westfield State University.
westfield.ma.edu/math/faculty/fleron/quotes/viewquote.asp?letter=p

POSITIVE ENGAGEMENT PROJECT. *Acing Math (One Deck at a Time): A Collection of Math Games,* The Positive Engagement Project website.
pepnonprofit.org/uploads/2/7/7/2/2772238/acing_math.pdf
pepnonprofit.org/mathematics.html

REULBACH, JULIE. "Math Games Collection on Google Docs—Add Your Game Today!" I Speak Math blog, November 18, 2013.
ispeakmath.org/2013/11/18/math-games-collection-on-google-docs-add-your-game-today

RONDA, ERLINA R. "Just like the games we play..." from "The Fun in Learning Mathematics Is in the Challenge," Mathematics for Teaching blog, November 12, 2011. Ronda is a mathematics education specialist at the University of the Philippines and also writes a math puzzle blog for students, K–12 Math Problems.
math4teaching.com/2011/11/12/fun-in-learning-mathematics-challenge
math-problems.math4teaching.com

SAWYER, W W. "Some time ago, I was told..." from "Abstract and Concrete," the transcription of a recording made for "Q.E.D.: A Series on the Teaching of Mathematics," August 11, 1961. (Edited slightly to correct an obvious transcription error.)
wwsawyer.org/pdf/abstract-and-concrete.pdf

SCARNE, JOHN, WITH CLAYTON RAWSON. *Scarne on Dice,* Military Service Publishing Co., 1945.

SOROOSHIAN, PAM. "Mathematicians don't sit around..." from the old Unschooling Discussion Yahoo group, quoted by Sandra Dodd in "Games and Math," Sandra Dodd's Unschoolers and Mathematics website.
sandradodd.com/math/pamgames

—. "Pam Sorooshian on Dice," Sandra Dodd's Unschoolers and Mathematics website.
sandradodd.com/math/pamdice

SULLIVAN, CHRISTINE. "My Closest Neighbor (Estimating with Fractions),"
A Sea of Math blog, August 4, 2014.
a-sea-of-math.blogspot.com/2014/08/my-closest-neighbor-estimating-with.html

SUN, ROBERT. 24 Game website.
24game.com

VANHATTUM, SUE. "Most people like games, so..." from *Playing with Math:
Stories from Math Circles, Homeschoolers, and Passionate Teachers,* Natural
Math, 2015. VanHattum is a community college mathematics teacher, math
circle leader, and blogger.
playingwithmath.org

WAKEFIELD, ALICE P. *Early Childhood Number Games: Teachers Reinvent Math
Instruction,* Allyn & Bacon, 1998.

WAY, JENNI. "Games can allow children to operate..." from "Learning
Mathematics Through Games Series: 1. Why Games?" Nrich Enriching
Mathematics website. Way is a math education researcher and associate professor at the University of Sydney.
nrich.maths.org/2489

WEDD, NICK, AND JOHN MCLEOD. "Mechanics of Card Games," Pagat website, May 15, 2009.
pagat.com/mech.html

WHITEHEAD, ALFRED NORTH. "From the very beginning of his education..."
from "The Aims of Education," in *The Aims of Education and Other Essays,*
Macmillan Company, 1929.
anthonyflood.com/whiteheadeducation.htm

WIKIPEDIA CONTRIBUTORS. "Pig (dice game)," Wikipedia Internet
Encyclopedia.
en.wikipedia.org/wiki/Pig_(dice_game)

ZASLAVSKY, CLAUDIA. "Language should be part of the activity..." from
*Preparing Young Children for Mathematics: A Book of Games with Updated
Book, Game and Resource Lists,* Schocken Books, 1986. Any book by Zaslavsky
is well worth reading.

—. *Tic Tac Toe and Other Three-in-a-Row Games from Ancient Egypt to the
Modern Computer,* Thomas Y. Crowell, 1982..

Index

About
the Author

DENISE GASKINS ENJOYS MATH, AND she delights in sharing that joy with young people. "Math is not just rules and rote memory," she says. "Math is like ice cream, with more flavors than you can imagine. And if all you ever do is textbook math, that's like eating broccoli-flavored ice cream."

A veteran homeschooling mother of five, Denise has taught or tutored mathematics at every level from pre-K to undergraduate physics. "Which," she explains, "at least in the recitation class I taught, was just one story problem after another. What fun!"

Now she writes the popular blog Let's Play Math and manages the Math Teachers at Play monthly math education blog carnival.

A Note from Denise

I hope you enjoyed this Math You Can Play book and found new ideas that will help your children enjoy learning.

If you believe these math games are worth sharing, please consider posting a review at Goodreads.com or at your favorite bookseller's website. Just a few lines would be great. An honest review is the highest compliment you can pay to any author, and your comments help fellow readers discover good books.

Thank you!

—DENISE GASKINS
LETSPLAYMATH@GMAIL.COM

Let's Connect Online

LET'S PLAY MATH BLOG
DeniseGaskins.com

FACEBOOK PAGE
facebook.com/letsplaymath

TWITTER
twitter.com/letsplaymath

GOOGLE+
plus.google.com/+DeniseGaskins

PINTEREST
pinterest.com/denisegaskins

EMAIL
LetsPlayMath@gmail.com

Playful Family Math

A Facebook Discussion Group

Want to help your kids learn math and enjoy it? Check out my new Facebook discussion group, where you can ask questions, share articles about learning math, or tell us your favorite math games, books and resources. This is a positive, supportive group for parents and teachers—and grandparents, aunts and uncles, caregivers, or anyone else—interested in talking about math concepts and creative ways to help children learn. Let's make math a playful family adventure!

facebook.com/groups/playfulfamilymath

Or get a free copy of my 24-page problem-solving booklet when you sign up for my newsletter mailing list. Plus you'll be among the first to hear about new books, revisions, and sales or other promotions.

TabletopAcademy.net/Subscribe

Books by Denise Gaskins

Let's Play Math:
How Families Can Learn Math Together—and Enjoy It

The *Math You Can Play* Series:
Counting & Number Bonds
Addition & Subtraction
Math You Can Play Combo
Multiplication & Fractions
Prealgebra & Geometry (upcoming)

Reviews

"In a culture where maths anxiety is now a diagnosable problem, this book shows the way to maths joy."

"With this approach I can teach my kids to think like mathematicians without worrying about leaving gaps."

"There were so many parts of this book that I highlighted that I really gave my Kindle a workout!"

"Most of the games can easily be scaled up for older kids, teens, and even adults. These are not drills disguised as games, but activities that require problem solving and strategy as well as calculation.

Let's Play Math:

How Families Can Learn Math Together

—and Enjoy It

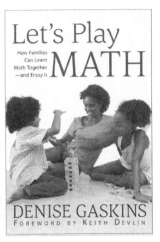

ALL PARENTS AND TEACHERS HAVE one thing in common: we want our children to understand and be able to use math. Filled with stories and pictures, *Let's Play Math* offers a wealth of practical, hands-on ideas for exploring math concepts from preschool to high school.

Your children will gain a strong foundation when you approach math as a family game, playing with ideas. Sections include:

HOW TO UNDERSTAND MATH: Introduce your children to the thrill of conquering a challenge. Build deep understanding by thinking, playing, and asking questions like a mathematician.

PLAYFUL PROBLEM SOLVING: Awaken your children's minds to the beauty and wonder of mathematics. Discover the social side of math, and learn games for players of all ages.

MATH WITH LIVING BOOKS: See how mathematical ideas ebb and flow through the centuries with this brief tour through history. Can your kids solve math puzzles from China, India, or Ancient Egypt?

LET'S GET PRACTICAL: Fit math into your family's daily life, help your children develop mental calculation skills, and find out what to try when your child struggles with schoolwork.

RESOURCES AND REFERENCES: With these lists of library books and Internet sites, you'll never run out of playful math to explore.

Denise Gaskins provides a treasure trove of helpful tips for all families, whether your children are homeschooling, unschooling, or attending a traditional classroom. Even if you struggled with math in school, you can help your kids practice mental math skills, master the basic facts, and ask the kind of questions that encourage deeper thought.

Don't let your children suffer from the epidemic of math anxiety. Grab a copy of *Let's Play Math*, and start enjoying math today.

The *Math You Can Play* Series

ARE YOU TIRED OF THE daily homework drama? Do your children sigh, fidget, whine, stare out the window—anything except work on their math? Wouldn't it be wonderful if math were something your kids *wanted* to do?

With the *Math You Can Play* series, your children can practice their math skills by playing games with basic items you already have around the house, such as playing cards and dice.

Math games pump up mental muscle, reduce the fear of failure, and develop a positive attitude toward mathematics. Through playful interaction, games strengthen a child's intuitive understanding of numbers and build problem-solving strategies. Mastering a math game can be hard work, but kids do it willingly because it's fun.

So what are you waiting for? Clear off a table, grab a deck of cards, and let's play some math!

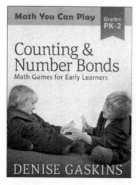

Counting & Number Bonds: Math Games for Early Learners

Preschool to Second Grade: Young children can play with counting and number recognition, while older students explore place value, build number sense, and begin learning the basics of addition.

Addition & Subtraction: Math Games for Elementary Students

Kindergarten to Fourth Grade: Children develop mental flexibility by playing with numbers, from basic math facts to the hundreds and beyond. Logic games build strategic thinking skills, and dice games give students hands-on experience with probability.

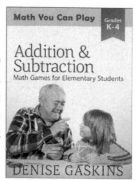

Math You Can Play Combo:
Number Games for Young Learners

Preschool to Fourth Grade: A combined volume, two books in one, with 42 kid-tested games that offer a variety of challenges for preschool and school-age learners. Help your children master the math facts and build a foundation for future learning.

Multiplication & Fractions:
Math Games for Tough Topics

Second to Sixth Grade: Students learn several math models that provide a sturdy foundation for understanding multiplication and fractions. The games feature times table facts and more advanced concepts such as division, fractions, decimals, and multistep mental math.

Prealgebra & Geometry:
Math Games for Middle School

Fourth to Ninth Grade: (planned for 2018) Older students can handle more challenging games that develop logic and problem-solving skills. Here are playful ways to explore positive and negative integers, number properties, mixed operations, functions, and coordinate geometry.

Fantasy Novels by Teresa Gaskins

The Riddled Stone Series

Banished

Hunted

Betrayed

Reviews

"A captivating fantasy story with a well-thought-out plot that would be a credit to any writer. But it is especially remarkable coming from a thirteen-year-old student who has been homeschooled all her life."

"People who like medieval-style fantasies with wraiths, spirits, and even an attacking swamp tree will enjoy the story. The excitement, adventure, and suspense will easily keep the reader's attention."

"The setting is a world of 'light' magic. Magic is rare, constrained, and follows a sort of logic, which may or not be fully understood by the people in the world. I like the way in which this sets up plot connections and forces things to happen for a reason, rather than deus ex machina or authorial patronus."

Banished: Who Stole the Magic Shard?

All Christopher Fredrico wanted was to be a peaceful scholar who could spend a lot of time with his friends. Now, falsely accused of stealing a magical artifact, Chris is forced to leave the only home he knows.

But as he and his friends travel towards the coast, they find a riddle that may save a kingdom—or cost them their lives.

Hunted: Magic is a Dangerous Guide

As a child, Terrin of Xell was almost devoured by a spirit from the Dark Forest. She knows better than to trust magic. But when her friend Chris was accused of a magical crime he didn't commit, she couldn't let him face banishment alone.

So she and her friends get caught up in a quest to recover an ancient relic, with only magic to guide them. And everything is going wrong.

Betrayed: How Can a Knight Fight Magic?

Trained by the greatest knight in North Raec, Sir Arnold Fredrico dreamed of valiant deeds. Save the damsel. Serve the king.

Dreams change. Now the land teeters at the brink of war. As a fugitive with a price on his head, Arnold struggles to protect his friends.

But his enemy wields more power than the young knight can imagine.

Made in the USA
San Bernardino, CA
06 April 2018